Student Solutions Manual

to accompany

Mathematics
Beyond
the Numbers

George T. Gilbert
Texas Christian University

Rhonda L. Hatcher
Texas Christian University

John Wiley & Sons, Inc.
New York • Chichester • Weinheim • Brisbane • Singapore • Toronto

ISBN 0-471-29397-0

Printed in the United States of America

10 9 8 7 6 5 4 3 2 1

Printed and bound by Hamilton Printing Company

CONTENTS

CHAPTER 1
Student Solution Manual

SECTION 1.1

1. The vote tally is Chinese 3, French 1, Indian 1, Italian 2, and Mexican 2, so they will choose Chinese.

3. **(a)** Let x be the number of votes Flores needs to ensure at least a tie for first place. If Payne gets all the votes that do not go to Flores and the race ends with a tie between Flores and Payne, then

$$31 + x = 23 + (30 - x).$$

Solving for x, we get

$$31 + x = 53 - x$$
$$x + x = 53 - 31$$
$$2x = 22$$
$$x = \tfrac{22}{2} = 11.$$

Therefore, to be assured of a win Flores needs 12 votes.

(b) Let x be the number of votes Payne needs to ensure at least a tie for first place. If Flores gets all the votes that do not go to Payne and the race ends with a tie between Payne and Flores, we have

$$23 + x = 31 + (30 - x).$$

Solving for x gives

$$23 + x = 61 - x$$
$$x + x = 61 - 23$$
$$2x = 38$$
$$x = \tfrac{38}{2} = 19.$$

Therefore, Payne needs 20 votes to ensure a win.

(c) Let x be the number of votes Bronowski needs to ensure at least a tie for first place. If Flores gets all the votes that do not go to Bronowski and the race ends with a tie between Bronowski and Flores, then

$$16 + x = 31 + (30 - x).$$

Solving for x gives

$$16 + x = 61 - x$$
$$x + x = 61 - 16$$
$$2x = 45$$
$$x = \tfrac{45}{2} = 22.5.$$

Therefore, Bronowski needs 23 votes to ensure a win.

5. Dividing 302 by 5 we get 60 with 2 left over. Therefore, some candidate must receive at least 61 votes. However, if a candidate has 61 votes, then some other candidate must also have at least 61 votes because otherwise at most $61 + 4 \cdot 60 = 301$ votes are cast. Thus, the winner must have at least 62 votes. One possible tally is 62, 60, 60, 60, 60.

7. **(a)** A total of $40{,}856 + 13{,}053 = 53{,}909$ voters voted for Crawford or miscellaneous candidates. If these voters voted for Jackson, he would again win a plurality of the votes. If they voted for Adams, then Adams would win a plurality with a total of $113{,}122 + 53{,}909 = 167{,}031$ votes to Jackson's $151{,}271$ votes and Clay's $47{,}531$ votes. If they voted for Clay, he would have a total of $47{,}531 + 53{,}909 = 101{,}440$ votes, not enough to beat Jackson. Therefore, only Jackson or Adams could have won a plurality of the votes.

(b) Let x be the number of votes Jackson would need to be assured of at least tie for first place. If Adams receives all the votes that do not go to Jackson and the election ends in a tie between Jackson and Adams, then

$$151{,}271 + x = 113{,}122 + (53{,}909 - x)$$

Solving for x, we get

$$151{,}271 + x = 167{,}031 - x$$
$$x + x = 167{,}031 - 151{,}271$$
$$2x = 15{,}760$$
$$x = \frac{15{,}760}{2} = 7880.$$

Therefore, Jackson needs 7881 votes or $\frac{7881}{53{,}909} \approx 0.1462 = 14.62\%$ of the Crawford and miscellaneous candidates voters.

9. The total number of votes cast was

$$64 + 62 + 38 = 164.$$

A majority is given by

$$\tfrac{1}{2}(164) + 1 = 83 \text{ votes,}$$

and therefore Clarke would have needed

$$83 - 64 = 19$$

of the Redwood supporters to win the final runoff. This is $\frac{19}{38} = 0.50 = 50\%$ of Redwood supporters.

11. **(a)** The vote tally would be

carpet	4
ceramic tile	2
wood	3

so carpet would win.

(b) The vote tally in the runoff would be

carpet	4
wood	$3 + 2 = 5$

so wood would win.

13. **(a)** The vote tally would be

10 p.m.	9%
12 midnight	28%
1 a.m.	31%
no curfew	32%

so no curfew would win.

(b) The vote tally in the runoff would be

$$1 \text{ a.m.} \qquad 31\% + 37\% = 68\%$$
$$\text{no curfew} \quad 32\%$$

so 1 a.m. would win.

15. **(a)** The vote tally would be

0 hours	3
5 hours	2
10 hours	4
20 hours	2
40 hours	8

so 40 hours would win.

(b) The vote tally in the runoff would be

10 hours	$4 + 5 = 9$
40 hours	$8 + 2 = 10$

so 40 hours would win.

(c) No, no matter how the three voters voted none of the contenders could gain more than 3 votes. Therefore, no alternate choice could obtain as many as the 8 votes assured to 40 hours.

(d) Yes, they could vote for 20 hours on the first ballot resulting in a first ballot vote tally of

0 hours	0
5 hours	2
10 hours	4
20 hours	5
40 hours	8.

In the runoff between 20 hours and 40 hours the vote tally would be

10 hours	$5 + 6 = 11$
40 hours	8

so 20 hours would win.

17. **(a)** The vote tally would be

Football	2
Soccer	1
Softball	4
Volleyball	5

so volleyball would win.

(b) Yes, they could vote for softball resulting in a vote tally of

Football	0
Soccer	1
Softball	6
Volleyball	5

so softball would win.

19. The voters with first choice softball would vote for their second choices giving a vote tally of

Football	$2 + 1 = 3$
Soccer	$1 + 3 = 4$
Volleyball	5

so volleyball would win.

21. In a contest between only two choices, each voter will vote for the choice he or she ranked higher. The results of the first vote will be football 5 and soccer 7. The results of the second vote will be soccer 4 and softball 8. The results of the third and final vote will be softball 7 and volleyball 5. Therefore, they will choose softball.

23. Yes, one way is to decide between softball and volleyball, then between the winner of that vote and football, then between the winner of that vote and soccer. The results of the first vote will be softball 7 and volleyball 5. The results of the second vote will be softball 5 and football 7. The results of the third and final vote will be football 5 and soccer 7. Therefore, soccer will be the winner.

25. In the first round of the vote, football will beat volleyball by 7 votes to 5, and softball will beat soccer by 8 votes to 4. The results of the second round vote will be football 7 and softball 5. Therefore, football wins.

SECTION 1.2

1. The Borda counts are

$$
\begin{array}{ll}
\text{Cardona} & 6\cdot4 + 2\cdot3 + 4\cdot2 + 2\cdot1 = 40 \\
\text{Pitts-Jones} & 4\cdot4 + 3\cdot3 + 6\cdot2 + 1\cdot1 = 38 \\
\text{De Plata} & 2\cdot4 + 8\cdot3 + 1\cdot2 + 3\cdot1 = 37 \\
\text{Vincent} & 2\cdot4 + 1\cdot3 + 3\cdot2 + 8\cdot1 = 25
\end{array}
$$

so Cardona was the winner.

3. The Borda counts are

$$
\begin{array}{ll}
\text{Bears} & 1\cdot5 + 2\cdot4 + 2\cdot3 + 1\cdot2 + 2\cdot1 = 23 \\
\text{Falcons} & 2\cdot5 + 1\cdot4 + 2\cdot3 + 3\cdot2 + 0\cdot1 = 26 \\
\text{Trojans} & 2\cdot5 + 1\cdot4 + 1\cdot3 + 2\cdot2 + 2\cdot1 = 23 \\
\text{Mustangs} & 1\cdot5 + 1\cdot4 + 1\cdot3 + 2\cdot2 + 3\cdot1 = 19 \\
\text{Horned Frogs} & 2\cdot5 + 3\cdot4 + 2\cdot3 + 0\cdot2 + 1\cdot1 = 29
\end{array}
$$

so Horned Frogs is the winner.

5. The Borda counts are

$$
\begin{array}{ll}
\text{Monday} & 18\cdot3 + 12\cdot2 + 15\cdot1 = 93 \\
\text{Wednesday} & 15\cdot3 + 16\cdot2 + 14\cdot1 = 91 \\
\text{Thursday} & 12\cdot3 + 17\cdot2 + 16\cdot1 = 86
\end{array}
$$

so Monday was the winner.

7. The Borda counts are

$$
\begin{array}{ll}
\text{Kahn} & 39\cdot3 + 23\cdot2 + 38\cdot1 = 201 \\
\text{Papini} & 27\cdot3 + 42\cdot2 + 31\cdot1 = 196 \\
\text{Tilson} & 34\cdot3 + 35\cdot2 + 31\cdot1 = 203
\end{array}
$$

so Tilson would have been the winner.

9. (a) The Borda counts are

$$
\begin{array}{ll}
\text{Tom Brokaw} & 6\cdot3 + 11\cdot2 + 5\cdot1 = 45 \\
\text{Peter Jennings} & 8\cdot3 + 6\cdot2 + 8\cdot1 = 44 \\
\text{Dan Rather} & 8\cdot3 + 5\cdot2 + 9\cdot1 = 43
\end{array}
$$

so Tom Brokaw has the top Borda count.

(b)   The tally would be

$$
\begin{array}{ll}
\text{Tom Brokaw} & 6 \\
\text{Peter Jennings} & 8 \\
\text{Dan Rather} & 8
\end{array}
$$

so Peter Jennings and Dan Rather tie for first place in a plurality vote.

(c)   The vote tally in the runoff would be

$$
\begin{array}{ll}
\text{Peter Jennings} & 8 + 3 = 11 \\
\text{Dan Rather} & 8 + 3 = 11
\end{array}
$$

so Peter Jennings and Dan Rather again tie.

11. **(a)**   The vote tally is

$$
\begin{array}{ll}
\text{Soccer} & 7 \\
\text{Baseball} & 8 \\
\text{Football} & 10
\end{array}
$$

so football wins.

(b)   The vote tally in the runoff is

$$
\begin{array}{ll}
\text{Soccer} & 8 + 4 = 12 \\
\text{Football} & 10 + 3 = 13
\end{array}
$$

so football wins.

(c)   The Borda counts are

$$
\begin{array}{ll}
\text{Soccer} & 7 \cdot 3 + 11 \cdot 2 + 7 \cdot 1 = 50 \\
\text{Baseball} & 8 \cdot 3 + 8 \cdot 2 + 9 \cdot 1 = 49 \\
\text{Football} & 10 \cdot 3 + 6 \cdot 2 + 9 \cdot 1 = 51
\end{array}
$$

so football wins.

(d)   Yes, if they ranked baseball first, soccer second, and football third then the Borda counts would be

$$
\begin{array}{ll}
\text{Soccer} & 50 - 4 = 46 \\
\text{Baseball} & 49 + 4 = 53 \\
\text{Football} & 51
\end{array}
$$

and baseball would win.

13. **(a)**   The Borda counts are

$$
\begin{array}{ll}
\text{Le Breton} & 1 \cdot 6 + 1 \cdot 5 + 2 \cdot 4 + 1 \cdot 3 + 2 \cdot 2 + 3 \cdot 1 = 29 \\
\text{Frye} & 4 \cdot 6 + 0 \cdot 5 + 0 \cdot 4 + 1 \cdot 3 + 2 \cdot 2 + 3 \cdot 1 = 34 \\
\text{Reeves} & 1 \cdot 6 + 3 \cdot 5 + 0 \cdot 4 + 4 \cdot 3 + 0 \cdot 2 + 2 \cdot 1 = 35 \\
\text{Keen-Sims} & 2 \cdot 6 + 0 \cdot 5 + 2 \cdot 4 + 1 \cdot 3 + 4 \cdot 2 + 1 \cdot 1 = 32 \\
\text{Cullors} & 0 \cdot 6 + 5 \cdot 5 + 4 \cdot 4 + 0 \cdot 3 + 0 \cdot 2 + 1 \cdot 1 = 42 \\
\text{Allen} & 2 \cdot 6 + 1 \cdot 5 + 2 \cdot 4 + 3 \cdot 3 + 2 \cdot 2 + 0 \cdot 1 = 38
\end{array}
$$

so Cullors has the top Borda count.

(b)   Yes, they could vote so that Allen wins. One way they could do this is by ranking Allen first, Le Breton second, Keen-Sims third, Reeves fourth, Cullors fifth, and Frye sixth. The Borda counts would then be

$$\begin{array}{ll} \text{Le Breton} & 29 + 2 = 31 \\ \text{Frye} & 34 \\ \text{Reeves} & 35 \\ \text{Keen-Sims} & 32 + 2 \cdot 2 = 36 \\ \text{Cullors} & 42 - 2 \cdot 3 = 36 \\ \text{Allen} & 38. \end{array}$$

(c) The member ranked Cullors third, so the only outcomes the member would prefer to Cullors are Keen-Sims and Reeves. The most the member can lower Cullors's Borda count is from 42 to 39 by ranking Cullors sixth. The member could not raise the Borda count of Keen-Sims above 32 because the member originally ranked Keen-Sims first. The member could raise the Borda count of Reeves by at most 1 point to 36 by ranking Reeves first. Therefore, the member could not make either Keen-Sims or Reeves beat Cullors and thus could not obtain a preferable result.

15. In this alternate method each candidate receives one less point from each voter. Therefore, if there are v voters, each candidate's count under this alternate method is exactly v points less than the Borda count as defined in the text. Therefore, the candidate with the most points is always the winner of the Borda count according to the definition in the text.

SECTION 1.3

1. (a) The vote tally is Bauer 5 and Sanders 7, so Sanders is the winner.

   (b) The vote tally is Bauer 7 and Donevska 5, so Bauer is the winner.

   (c) The vote tally is Donevska 4 and Sanders 8, so Sanders is the winner.

   (d) Sanders is the Condorcet winner.

3. Mexican beats Cajun 15 to 6, Mexican beats Chinese 18 to 3, Mexican beats Indian 20 to 1, and Mexican beats Italian 16 to 5. Therefore, Mexican is the Condorcet winner.

5. Julia is defeated by Wolfgang 5 to 2, Wolfgang is defeated by Paul 4 to 3, and Paul is defeated by Julia 4 to 3. Therefore, there is no Condorcet winner.

7. Hungry Boar beats Hungry Dog 58% to 42% and Hungry Boar beats Hungry Pig 51% to 49%, so Hungry Boar is the Condorcet winner.

9. (a) The Borda counts are

   $$\begin{array}{ll} \text{Mandatory uniforms} & 41 \cdot 3 + 17 \cdot 2 + 61 \cdot 1 = 218 \\ \text{Optional uniforms} & 35 \cdot 3 + 78 \cdot 2 + 6 \cdot 1 = 267 \\ \text{No uniforms} & 43 \cdot 3 + 24 \cdot 2 + 52 \cdot 1 = 229 \end{array}$$

   so optional uniforms has the top Borda count.

   (b) The vote tally would be

   $$\begin{array}{ll} \text{Mandatory uniforms} & 41 \\ \text{Optional uniforms} & 35 \\ \text{No uniforms} & 43 \end{array}$$

   so no uniforms would win a plurality of the vote.

   (c) The vote tally in the runoff would be

   $$\begin{array}{ll} \text{Mandatory uniforms} & 41 + 16 = 57 \\ \text{No uniforms} & 43 + 19 = 62 \end{array}$$

   so no uniforms would win.

(d) Optional uniforms beats mandatory uniforms 77 to 42 and optional uniforms beats no uniforms 71 to 48. Therefore, optional uniforms is the Condorcet winner.

11. **(a)** The Borda counts are

Beagle	$3 \cdot 4 + 0 \cdot 3 + 2 \cdot 2 + 1 \cdot 1 = 17$
Fox terrier	$1 \cdot 4 + 1 \cdot 3 + 3 \cdot 2 + 1 \cdot 1 = 14$
Bulldog	$2 \cdot 4 + 2 \cdot 3 + 0 \cdot 2 + 2 \cdot 1 = 16$
Boxer	$0 \cdot 4 + 3 \cdot 3 + 1 \cdot 2 + 2 \cdot 1 = 13$

so beagle has the top Borda count.

(b) The vote tally would be

Beagle	3
Fox terrier	1
Bulldog	2
Boxer	0

so beagle would win a plurality of the vote.

(c) The vote tally in the runoff would be

Beagle	$3 + 1 = 4$
Bulldog	2

so beagle would win.

(d) Beagle and fox terrier tie 3 to 3, bulldog is defeated by beagle 4 to 2, and boxer is defeated by beagle 4 to 2. Therefore, there is no Condorcet winner. (However, beagle is a weak Condorcet winner.)

13. No, they are not single-peaked with respect to the ordering of candidates beagle, fox terrier, bulldog, boxer. If the preference rankings were single-peaked with respect to this order, then any voter ranking beagle first would have to rank fox terrier second, bulldog third, and boxer fourth. Notice that this is not the case for any of the three voters who ranked beagle first.

15. **(a)** The vote tally would be

private development	27
public development	30
recommitment	18

so public development would win.

(b) The vote tally in the runoff would be

private development	$27 + 16 = 43$
public development	$30 + 2 = 32$

so private development would win.

(c) The Borda counts are

private development	$27 \cdot 3 + 20 \cdot 2 + 28 \cdot 1 = 149$
public development	$30 \cdot 3 + 11 \cdot 2 + 34 \cdot 1 = 146$
recommitment	$18 \cdot 3 + 44 \cdot 2 + 13 \cdot 1 = 155$

so recommitment is the Borda winner.

(d) Private development is defeated by recommitment 44 to 31, recommitment is defeated by public development 39 to 36, and public development is defeated by private development 43 to 32. Therefore, there is no Condorcet winner.

17. We would expect the choices in (a) and (b) to have single-peaked preferences among those voting.

19. Yes, the candidate must win a plurality election with a runoff between the top two finishers because under the assumption that there are no ties for the runoff position we know the runoff will be a head-to-head comparison between the candidate and one other candidate. Because the candidate is the Condorcet winner, we are assured that the candidate will win this head-to-head comparison.

21. In a head-to-head comparison between two candidates adjacent to one another in the ordering, all voters whose first choice is to the left of the two candidates will vote for the leftmost of the two, and voters whose first choice is to the right of the two candidates will vote for the rightmost of the two. Furthermore, if a candidate defeats an adjacent candidate, then the candidate will also defeat all candidates on the far side of the adjacent candidate. Thus, to find the Condorcet winner, we can move from left to right along the ordering of candidates, keeping a running tally of first place votes. The candidate whose first place votes take the running tally from less than 50% of the first place votes to over 50% is the Condorcet winner because this candidate will beat both adjacent candidates head-to-head and therefore all other candidates head-to-head. If all of a candidate's first place votes take the running tally to exactly 50%, then this candidate through the next candidate to receive any first place votes are weak Condorcet winners because they will tie in head-to-head comparisons and beat all other candidates head-to-head.

SECTION 1.4

1. The vote tally is

McClain	5
Snyder	6
Freeman	7
Sanders	3
Yang	4

so Freeman is the club's new president.

3. **(a)** The vote tally is

A	28,073
B	19,753
C	11,221
D	23,992

so A is the winner.

(b) The 523 engineers who approved of all four candidates had no potential to help determine the winner. Because there were 54,727 voters in all, these 523 engineers were

$$\frac{523}{54{,}727} \approx 0.00956 = 0.956\%$$

of the voters.

5. The vote tally would be

De Castro	48%
Telger	47%
Segura	45%

so De Castro would win an approval election.

7. **(a)** The plurality vote tally would be

Sixteen	0
Seventeen	0
Eighteen	12
Nineteen	0
Twenty	0
Twenty-one	4

so eighteen wins.

(b) Because eighteen won a majority of votes in the plurality vote, it would also win a plurality vote with a runoff between the top two finishers.

(c) The Borda counts are

Sixteen	$0{\cdot}6 + 0{\cdot}5 + 0{\cdot}4 + 1{\cdot}3 + 2{\cdot}2 + 13{\cdot}1 = 20$
Seventeen	$0{\cdot}6 + 2{\cdot}5 + 3{\cdot}4 + 0{\cdot}3 + 11{\cdot}2 + 0{\cdot}1 = 44$
Eighteen	$12{\cdot}6 + 0{\cdot}5 + 0{\cdot}4 + 4{\cdot}3 + 0{\cdot}2 + 0{\cdot}1 = 84$
Nineteen	$0{\cdot}6 + 10{\cdot}5 + 6{\cdot}4 + 0{\cdot}3 + 0{\cdot}2 + 0{\cdot}1 = 74$
Twenty	$0{\cdot}6 + 4{\cdot}5 + 7{\cdot}4 + 4{\cdot}3 + 0{\cdot}2 + 1{\cdot}1 = 61$
Twenty-one	$4{\cdot}6 + 0{\cdot}5 + 0{\cdot}4 + 7{\cdot}3 + 3{\cdot}2 + 2{\cdot}1 = 53$

so eighteen wins the Borda count.

(d) Because eighteen won a majority of the votes in the plurality vote, it would beat every other candidate in head-to-head comparisons.

(e) The approval vote tally would be

Sixteen	0
Seventeen	3
Eighteen	12
Nineteen	7
Twenty	3
Twenty-one	4

so eighteen wins.

9. (a) The plurality vote tally is

Douglas	0
Holmes	4
Cochrane	0
Mason	2
Bailey	5

so Bailey wins.

(b) The vote tally in the runoff is

Holmes	$4 + 2 = 6$
Bailey	5

so Holmes wins.

(c) The Borda counts are

Douglas	$0{\cdot}5 + 0{\cdot}4 + 0{\cdot}3 + 7{\cdot}2 + 4{\cdot}1 = 18$
Holmes	$4{\cdot}5 + 6{\cdot}4 + 1{\cdot}3 + 0{\cdot}2 + 0{\cdot}1 = 47$
Cochrane	$0{\cdot}5 + 0{\cdot}4 + 1{\cdot}3 + 4{\cdot}2 + 6{\cdot}1 = 17$
Mason	$2{\cdot}5 + 2{\cdot}4 + 7{\cdot}3 + 0{\cdot}2 + 0{\cdot}1 = 39$
Bailey	$5{\cdot}5 + 3{\cdot}4 + 2{\cdot}3 + 0{\cdot}2 + 1{\cdot}1 = 44$

so Holmes wins the Borda count.

(d)   Holmes beats Douglas 11 to 0, Holmes beats Cochrane 11 to 0, Holmes beats Mason 8 to 3, and Holmes beats Bailey 6 to 5. Therefore, Holmes is the Condorcet winner.

(e)   The approval vote tally is

Douglas	0
Holmes	6
Cochrane	1
Mason	5
Bailey	8

so Bailey wins.

(f)   No, the voter could not raise the approval vote tallies of Holmes or Mason, and the most the voter could lower Bailey's tally is by one. This would still leave Bailey with a tally of 7 which would beat Holmes and Mason.

11.   (a)   The plurality vote tally is

White	9
Green	5
Ivory	7

so white wins.

(b)   The vote tally in the runoff is

White	$9 + 1 = 10$
Ivory	$7 + 4 = 11$

so ivory wins.

(c)   The Borda counts are

White	$9{\cdot}3 + 5{\cdot}2 + 7{\cdot}1 = 44$
Green	$5{\cdot}3 + 7{\cdot}2 + 9{\cdot}1 = 38$
Ivory	$7{\cdot}3 + 9{\cdot}2 + 5{\cdot}1 = 44$

so white and ivory tie for first place.

(d)   Ivory beats white 11 to 10 and ivory beats green 12 to 9. Therefore, ivory is the Condorcet winner.

(e)   The approval vote tally is

White	10
Green	6
Ivory	11

so ivory wins.

(f)   Yes, if the two voters who approved of both white and ivory instead approved only of white, then the approval vote tally would be

White	10
Green	6
Ivory	9

and white would win.

13.   (a)   Reagan won a majority of the votes in the plurality vote, and therefore he would beat every other candidate in head-to-head comparisons.

(b)   Every voter approved of either one or two candidates. It is reasonable to assume that if a voter approved of one candidate, the candidate would be the voter's first choice, and that if a voter approved of two candidates, the candidates would be the voter's first and second choices. Because 51.6% of the people had Reagan as their first choice and 61% approved of him, it follows that $61\% - 51.6\% = 9.4\%$ of the people had Reagan as their second choice and approved of him.

(c) Because 41.7% of the people had Carter as their first choice and 57% approved of him, it follows that $57\% - 41.7\% = 15.3\%$ of the people had Carter as their second choice and approved of him.

(d) Because 6.7% of the people had Anderson as their first choice and 49% approved of him, we see that $49\% - 6.7\% = 42.3\%$ of the people had Anderson as their second choice and approved of him.

(e) The percentages from parts (b), (c), and (d) sum to $9.4\% + 15.3\% + 42.3\% = 67\%$. We estimate that $\frac{9.4\%}{67\%} \approx 14\%$ of the voters would have Reagan as their second choice, $\frac{15.3\%}{67\%} \approx 23\%$ would have Carter as their second choice, and $\frac{42.3\%}{67\%} \approx 63\%$ would have Anderson as their second choice.

(f) For each candidate, we can find the percentage of third place votes the candidate would receive by subtracting the percentage of first-place and second-place votes the candidate would receive from 100%. Using this, the plurality vote results, and our calculations from part (e), we find that the Borda counts are

$$
\begin{array}{lll}
\text{Reagan} & (51.6){\cdot}3 + (14){\cdot}2 + (34.4){\cdot}1 = 217.2 \\
\text{Carter} & (41.7){\cdot}3 + (23){\cdot}2 + (35.3){\cdot}1 = 206.4 \\
\text{Anderson} & (6.7){\cdot}3 + (63){\cdot}2 + (30.3){\cdot}1 = 176.4
\end{array}
$$

and Reagan would win.

15. Using the given information, we first compute the breakdown of the voters by their preference rankings.

$$
\begin{array}{l}
\text{DWRT: } 900{,}369 \\
\text{WDRT: } (0.10)(6{,}293{,}152) = 629{,}315.2 \approx 629{,}315 \\
\text{WTRD: } (0.20)(6{,}293{,}152) = 1{,}258{,}630.4 \approx 1{,}258{,}630 \\
\text{WRTD: } 6{,}293{,}152 - (629{,}315 + 1{,}258{,}630) = 4{,}405{,}207 \\
\text{RTWD: } (0.75)(4{,}119{,}207) = 3{,}089{,}405.25 \approx 3{,}089{,}405 \\
\text{RWTD: } 4{,}119{,}207 - 3{,}089{,}405 = 1{,}029{,}802 \\
\text{TRWD: } (0.80)(3{,}486{,}333) = 2{,}789{,}066.4 \approx 2{,}789{,}066 \\
\text{TWRD: } 3{,}486{,}333 - 2{,}789{,}066 = 697{,}267
\end{array}
$$

Therefore, we have the following table of preference rankings.

Number of Voters

	4,405,207	1,258,630	629,315	1,029,802	3,089,405	697,267	2,789,066	900,369
Wilson	1	1	1	2	3	2	3	2
Roosevelt	2	3	3	1	1	3	2	3
Taft	3	2	4	3	2	1	1	4
Debs	4	4	2	4	4	4	4	1

(a) If Roosevelt had not run, we assume that the voters who ranked him first would vote for their second choice and all other voters would vote for their first choice. The plurality vote tally would have been

$$
\begin{array}{ll}
\text{Wilson} & 6{,}293{,}152 + 1{,}029{,}802 = 7{,}322{,}954 \\
\text{Taft} & 3{,}486{,}333 + 3{,}089{,}405 = 6{,}575{,}738 \\
\text{Debs} & 900{,}369
\end{array}
$$

so Wilson would have won.

(b) If Taft was left out of the election, we assume that the voters who ranked him first would vote for their second choice and all other voters would vote for their first choice. The plurality vote tally would have been

$$
\begin{array}{ll}
\text{Wilson} & 6{,}293{,}152 + 697{,}267 = 6{,}990{,}419 \\
\text{Roosevelt} & 4{,}119{,}207 + 2{,}789{,}066 = 6{,}908{,}273 \\
\text{Debs} & 900{,}369
\end{array}
$$

so Wilson would have won.

(c)   The vote tally in a runoff between Wilson and Roosevelt would have been

$$\begin{array}{ll} \text{Wilson} & 6{,}293{,}152 + 1{,}597{,}636 = 7{,}890{,}788 \\ \text{Roosevelt} & 4{,}119{,}207 + 2{,}789{,}066 = 6{,}908{,}273 \end{array}$$

so Wilson would have won by a fairly small margin.

(d)   The Borda counts would have been

Wilson	$(6{,}293{,}152){\cdot}4 + (2{,}627{,}438){\cdot}3 + (5{,}878{,}471){\cdot}2 + 0{\cdot}1 = 44{,}811{,}846$
Roosevelt	$(4{,}119{,}207){\cdot}4 + (7{,}194{,}273){\cdot}3 + (3{,}485{,}581){\cdot}2 + 0{\cdot}1 = 45{,}030{,}809$
Taft	$(3{,}486{,}333){\cdot}4 + (4{,}348{,}035){\cdot}3 + (5{,}435{,}009){\cdot}2 + (1{,}529{,}684){\cdot}1 = 39{,}389{,}139$
Debs	$(900{,}369){\cdot}4 + (629{,}315){\cdot}3 + 0{\cdot}2 + (13{,}269{,}377) = 18{,}758{,}798$

so Roosevelt would have won.

(e)   Wilson beats Roosevelt 7,890,788 to 6,908,273, Wilson beats Taft 8,223,323 to 6,575,738, and Wilson beats Debs 13,898,692 to 900,369. Therefore, Wilson is a Condorcet winner.

(f)   The approval vote tallies would have been

Wilson	$6{,}293{,}152 + (0.25)(2{,}627{,}438) = 6{,}950{,}011.5 \approx 6{,}950{,}012$
Roosevelt	$4{,}119{,}207 + (0.25)(7{,}194{,}273) = 5{,}917{,}775.25 \approx 5{,}917{,}775$
Taft	$3{,}486{,}333 + (0.25)(4{,}348{,}035) = 4{,}573{,}341.75 \approx 4{,}573{,}342$
Debs	$900{,}369 + (0.25)(629{,}315) = 1{,}057{,}697.75 \approx 1{,}057{,}698$

so Wilson would have won.

17.   Using the given information, we first compute the breakdown of the voters by their preference rankings.

$$\begin{array}{l} \text{DWRT: } 900{,}369 \\ \text{WDRT: } (0.20)(6{,}293{,}152) = 1{,}258{,}630.4 \approx 1{,}258{,}630 \\ \text{WRTD: } 6{,}293{,}152 - 1{,}258{,}630 = 5{,}034{,}522 \\ \text{RTWD: } (0.85)(4{,}119{,}207) = 3{,}501{,}325.95 \approx 3{,}501{,}326 \\ \text{RWTD: } 4{,}119{,}207 - 3{,}501{,}326 = 617{,}881 \\ \text{TRWD: } (0.90)(3{,}486{,}333) = 3{,}137{,}699.7 \approx 3{,}137{,}700 \\ \text{TWRD: } 3{,}486{,}333 - 3{,}137{,}700 = 348{,}633 \end{array}$$

Therefore, we have the following table of preference rankings.

Number of Voters

	5,034,522	1,258,630	617,881	3,501,326	348,633	3,137,700	900,369
Wilson	1	1	2	3	2	3	2
Roosevelt	2	3	1	1	3	2	3
Taft	3	4	3	2	1	1	4
Debs	4	2	4	4	4	4	1

(a)   If Roosevelt had not run, the plurality vote tally would have been

$$\begin{array}{ll} \text{Wilson} & 6{,}293{,}152 + 617{,}881 = 6{,}911{,}033 \\ \text{Taft} & 3{,}486{,}333 + 3{,}501{,}326 = 6{,}987{,}659 \\ \text{Debs} & 900{,}369 \end{array}$$

so Taft would have won by a fairly small margin.

(b)   If Taft was left out of the election, the plurality vote tally would have been

$$\begin{array}{ll} \text{Wilson} & 6{,}293{,}152 + 348{,}633 = 6{,}641{,}785 \\ \text{Roosevelt} & 4{,}119{,}207 + 3{,}137{,}700 = 7{,}256{,}907 \\ \text{Debs} & 900{,}369 \end{array}$$

so Roosevelt would have won.

(c) The vote tally in a runoff between Wilson and Roosevelt would have been

$$\begin{array}{ll} \text{Wilson} & 6{,}293{,}152 + 1{,}249{,}002 = 7{,}542{,}154 \\ \text{Roosevelt} & 4{,}119{,}207 + 3{,}137{,}700 = 7{,}256{,}907 \end{array}$$

so Wilson would have won.

(d) The Borda counts would have been

$$\begin{array}{ll} \text{Wilson} & (6{,}293{,}152){\cdot}4 + (1{,}866{,}883){\cdot}3 + (6{,}639{,}026){\cdot}2 + 0{\cdot}1 = 44{,}051{,}309 \\ \text{Roosevelt} & (4{,}119{,}207){\cdot}4 + (8{,}172{,}222){\cdot}3 + (2{,}507{,}632){\cdot}2 + 0{\cdot}1 = 46{,}008{,}758 \\ \text{Taft} & (3{,}486{,}333){\cdot}4 + (3{,}501{,}326){\cdot}3 + (5{,}652{,}403){\cdot}2 + (2{,}158{,}999){\cdot}1 = 37{,}913{,}115 \\ \text{Debs} & (900{,}369){\cdot}4 + (1{,}258{,}630){\cdot}3 + 0{\cdot}2 + (12{,}640{,}062){\cdot}1 = 20{,}017{,}428 \end{array}$$

so Roosevelt would have won.

(e) Wilson beats Roosevelt 7,542,154 to 7,256,907, Wilson beats Taft 7,811,402 to 6,987,659, and Wilson beats Debs 13,898,692 to 900,369. Therefore, Wilson is a Condorcet winner.

(f) The approval vote tallies would have been

$$\begin{array}{ll} \text{Wilson} & 6{,}293{,}152 + (0.75)(1{,}866{,}883) = 7{,}693{,}314.25 \approx 7{,}693{,}314 \\ \text{Roosevelt} & 4{,}119{,}207 + (0.75)(8{,}172{,}222) = 10{,}248{,}373.5 \approx 10{,}248{,}374 \\ \text{Taft} & 3{,}486{,}333 + (0.75)(3{,}501{,}326) = 6{,}122{,}327.5 \approx 6{,}122{,}328 \\ \text{Debs} & 900{,}369 + (0.75)(1{,}258{,}630) = 1{,}844{,}341.5 \approx 1{,}844{,}341 \end{array}$$

so Roosevelt would have won.

19. We first compute the breakdown of the voters by their preference rankings. Among the Baldwin supporters, $(0.60)(78{,}264) = 46{,}958.4 \approx 46{,}958$ ranked Smith second and $78{,}264 - 46{,}958 = 31{,}306$ ranked Studley second. Similarly, $(0.70)(67{,}531) = 47{,}271.7 \approx 47{,}272$ Studley supporters ranked Smith second, while $67{,}531 - 47{,}272 = 20{,}259$ ranked Baldwin second. Finally, $(0.65)(31{,}020) = 20{,}163$ Smith supporters ranked Studley second, and $31{,}020 - 20{,}163 = 10{,}857$ ranked Baldwin second. Therefore, we have the following table of preference rankings.

Number of Voters

	31,306	46,958	20,259	47,272	10,857	20,163
Baldwin	1	1	2	3	2	3
Studley	2	3	1	1	3	2
Smith	3	2	3	2	1	1

(a) The vote tally in a runoff is

$$\begin{array}{ll} \text{Baldwin} & 78{,}264 + 10{,}857 = 89{,}121 \\ \text{Studley} & 67{,}531 + 20{,}163 = 87{,}694 \end{array}$$

so Baldwin is the winner.

(b) The Borda counts are

$$\begin{array}{ll} \text{Baldwin} & (78{,}264){\cdot}3 + (31{,}116){\cdot}2 + (67{,}435){\cdot}1 = 364{,}459 \\ \text{Studley} & (67{,}531){\cdot}3 + (51{,}469){\cdot}2 + (57{,}815){\cdot}1 = 363{,}346 \\ \text{Smith} & (31{,}020){\cdot}3 + (94{,}230){\cdot}2 + (51{,}565){\cdot}1 = 333{,}085 \end{array}$$

so Baldwin is the Borda winner.

(c) Baldwin beats Studley 89,121 to 87,694, and Baldwin beats Smith 98,523 to 78,292. Therefore, Baldwin is the Condorcet winner.

(d) The approval vote tallies are

$$
\begin{array}{ll}
\text{Baldwin} & 78{,}264 + (0.40)(31{,}116) = 90{,}710.4 \approx 90{,}710 \\
\text{Studley} & 67{,}531 + (0.40)(51{,}469) = 88{,}118.6 \approx 88{,}119 \\
\text{Smith} & 31{,}020 + (0.40)(94{,}230) = 68{,}712
\end{array}
$$

so Baldwin is the winner.

21. We first compute the breakdown of the voters by their preference rankings. We see that $(0.60)(37.5\%) = 22.5\%$ rank Ferris first and Watkins second, and $37.5\% - 22.5\% = 15\%$ rank Ferris first and Musselman second. Similarly, $(0.70)(32.9\%) = 23.03\%$ rank Musselman first and Watkins second, and $32.9\% - 23.03\% = 9.87\%$ rank Musselman first and Ferris second. Finally, $(0.65)(29.6\%) = 19.24\%$ rank Watkins first and Musselman second while $29.6\% - 19.24\% = 10.36\%$ rank Watkins first and Ferris second. Therefore, we have the following table of preference rankings.

Percentage of Voters

	15	22.5	9.87	23.03	10.36	19.24
Ferris	1	1	2	3	2	3
Musselman	2	3	1	1	3	2
Watkins	3	2	3	2	1	1

(a) The vote tally in a runoff is

$$
\begin{array}{ll}
\text{Ferris} & 37.5\% + 10.36\% = 47.86\% \\
\text{Musselman} & 32.9\% + 19.24\% = 52.14\%
\end{array}
$$

so Musselman is the winner.

(b) The Borda counts are

$$
\begin{array}{ll}
\text{Ferris} & (37.5)\cdot3 + (20.23)\cdot2 + (42.27)\cdot1 = 195.23 \\
\text{Musselman} & (32.9)\cdot3 + (34.24)\cdot2 + (32.86)\cdot1 = 200.04 \\
\text{Watkins} & (29.6)\cdot3 + (45.53)\cdot2 + (24.87)\cdot1 = 204.73
\end{array}
$$

so Watkins is the Borda winner.

(c) Watkins beats Ferris 52.63% to 47.37%, and Watkins beats Musselman 52.1% to 47.9%. Therefore, Watkins is the Condorcet winner.

(d) The approval vote tallies are

$$
\begin{array}{ll}
\text{Ferris} & 37.5\% + (0.40)(20.23\%) = 45.592\% \\
\text{Musselman} & 32.9\% + (0.40)(34.24\%) = 46.596\% \\
\text{Watkins} & 29.6\% + (0.40)(45.53\%) = 47.812\%
\end{array}
$$

so Watkins is the winner.

23. To compute the breakdown of the voters by their preference rankings we first note that $(0.50)(24\%) = 12\%$ rank Goodell first and Ottinger second and $(0.50)(24\%) = 12\%$ rank Goodell first and Buckley second. Using this and the assumption that preferences are single-peaked, with Goodell in the middle, we have the following table of preference rankings.

Percentage of Voters

	37%	12%	12%	39%
Ottinger	1	2	3	3
Goodell	2	1	1	2
Buckley	3	3	2	1

(a)   The vote tally in a runoff would be

$$\begin{aligned}
\text{Buckley} &\quad 39\% + 12\% = 51\% \\
\text{Ottinger} &\quad 37\% + 12\% = 49\%
\end{aligned}$$

so Buckley would win.

(b)   Goodell beats Ottinger 63% to 37%, and Goodell beats Buckley 61% to 39%. Therefore, Goodell is the Condorcet winner.

(c)   The approval vote tallies are

$$\begin{aligned}
\text{Ottinger} &\quad 37\% + (0.50)(12\%) = 43\% \\
\text{Goodell} &\quad 24\% + (0.50)(76\%) = 62\% \\
\text{Buckley} &\quad 39\% + (0.50)(12\%) = 45\%
\end{aligned}$$

so Goodell wins an approval vote.

25.   Assume the election has c candidates. By approving of x candidates and disapproving of $c - x$ candidates, a voter makes $x(c - x) = cx - x^2$ head-to-head distinctions. This is a downward parabola whose highest point occurs at $x = c/2$. You may see this by graphing $cx - x^2$ for between 0 and c. The value is largest for x as close as possible to $c/2$, i.e., $c/2$ for c even or $(c \pm 1)/2$ for c odd. Therefore, we see that the number of head-to-head distinctions is greatest when you approve of as close to half of all candidates as possible.

SECTION 1.5

1.   If every voter prefers candidate A to candidate B, then candidate B cannot get any first-place votes, so cannot be the plurality winner.

3.   We construct an example where candidate A barely defeats candidate B in a two-candidate race. We then bring in candidate C who will take enough points from A so that A falls below B, yet not enough so that C wins the election. One such example follows. (It is important to be sure that the first set of rankings is left when C is removed from the second set of rankings.)

Number of Voters		
	4	3
A	1	2
B	2	1

Number of Voters		
	4	3
A	1	3
B	2	1
C	3	2

In the two-candidate election, A has $4 \cdot 2 + 3 \cdot 1 = 11$ points to B's $3 \cdot 2 + 4 \cdot 1 = 10$ points. In the second, the Borda counts are

$$\begin{aligned}
\text{A} &\quad 4{\cdot}3 + 3{\cdot}1 = 15 \\
\text{B} &\quad 3{\cdot}3 + 4{\cdot}2 = 17 \\
\text{C} &\quad 3{\cdot}2 + 4{\cdot}1 = 10.
\end{aligned}$$

5.   Consider an election that initially involves two candidates, A and B. Suppose a third candidate, C, enters the race and that C is the third choice of all voters. Because there are now three candidates, those voters who rank A first, might approve of both A and B, whereas they might approve of only A in the two-candidate race between A and B. Those who rank B first, could approve only of B in both a two-candidate and three-candidate race. Combining these possibilities we turn a close victory for A into a comfortable victory for B.

	Number of Voters	
	4	3
A	1√	2
B	2	1√

	Number of Voters	
	4	3
A	1√	2
B	2√	1√
C	3	3

With just A and B, A gets 4 votes, B 3. With all three, the approval tally is A 4, B 7, C 0.

7. (a) The vote of each Representative or Senator is counted the same — as one yes or one no, so the method is anonymous.

(b) We give an example for the Senate, with 100 members. Assume amendments 1 and 2 apply to the same part of the bill, so that one replaces the other.

	Number of Voters		
	32	33	35
Bill with no amendment	1	3	4
Bill with amendment 1	2	4	2
Bill with amendment 2	3	1	3
No bill	4	2	1

If amendment 1 is voted on first, it will be defeated 65-35. Amendment 2 would then pass 68-32. The final vote on the bill with amendment 2 would pass 65-35. However, if amendment 2 is voted on first, it would pass 68-32. Amendment 1 would then replace amendment 2 by a 67-33 vote. Finally, the bill with amendment 2 would be defeated 68-32. Therefore, the method is not neutral.

(c) Suppose a bill passed the House 225-210 and the Senate 55-45. If 10 "no" voters from the House switched places with 10 "yes" voters from the Senate, the bill would pass the House 235-200, but lose 55-45 in the Senate. Thus, the procedure is not anonymous when all representatives and senators are taken into account.

9. (a) If every voter prefers A to B, then A defeats B head-to-head and B cannot be a Condorcet winner (or a weak Condorcet winner).

(b) Suppose A is the Condorcet winner in a race with B and perhaps other candidates. If a new candidate C enters the race, then A will still defeat B head-to-head, so that B cannot emerge as the Condorcet winner.

11. If one or more voters raises A in the rankings and keeps the relative rankings of other candidates the same, then A's Borda count will increase and the Borda counts of the other candidate will either decrease or remain unchanged. In either case, it is impossible for the result of the election to change from a win for A to a win for some other candidate.

CHAPTER REVIEW EXERCISES

1. Dividing 6 into 454 we get 75 with 4 left over. Therefore, some candidate must receive at least 76 votes. However, if a candidate has 76 votes, then some other candidate must also receive at least 76 votes because otherwise at most $76 + 5 \cdot 75 = 451$ votes are cast. Therefore, the winner must have at least 77 votes. One possible tally is 77, 76, 76, 75, 75, 75.

3. (a) The plurality vote tally would be

$$\begin{array}{ll} \text{Apples} & 4 \\ \text{Bananas} & 10 \\ \text{Grapes} & 9 \\ \text{Oranges} & 1 \end{array}$$

so bananas would win.

(b)   The vote tally in the runoff would be

$$\begin{array}{ll} \text{Bananas} & 10 + 3 = 13 \\ \text{Grapes} & 9 + 2 = 11 \end{array}$$

so bananas would win.

(c)   The Borda counts are

$$\begin{array}{ll} \text{Apples} & 4 \cdot 4 + 4 \cdot 3 + 9 \cdot 2 + 7 \cdot 1 = 53 \\ \text{Bananas} & 10 \cdot 4 + 6 \cdot 3 + 4 \cdot 2 + 4 \cdot 1 = 70 \\ \text{Grapes} & 9 \cdot 4 + 7 \cdot 3 + 6 \cdot 2 + 2 \cdot 1 = 71 \\ \text{Oranges} & 1 \cdot 4 + 7 \cdot 3 + 5 \cdot 2 + 11 \cdot 1 = 46 \end{array}$$

so grapes has the top Borda count.

(d)   Bananas beats apples 15 to 9, bananas beats grapes 13 to 11, and bananas beats oranges 18 to 6. Therefore, bananas would be the Condorcet winner.

5.   The vote tally is

$$\begin{array}{ll} \text{Comedian} & 6 \\ \text{Jazz trio} & 7 \\ \text{Piano} & 3 \\ \text{Rock band} & 5 \\ \text{Classical guitarist} & 6 \end{array}$$

so jazz trio wins.

7.   **(a)**   The plurality vote tally would be

$$\begin{array}{ll} \text{Perella} & 2 \\ \text{Mintz} & 3 \\ \text{Zukoff} & 3 \end{array}$$

so Mintz and Zukoff tie for first place.

(b)   The vote tally in the runoff would be

$$\begin{array}{ll} \text{Mintz} & 3 \\ \text{Zukoff} & 3 + 2 = 5 \end{array}$$

so Zukoff would win.

(c)   The Borda counts would be

$$\begin{array}{ll} \text{Perella} & 2 \cdot 3 + 1 \cdot 2 + 5 \cdot 1 = 13 \\ \text{Mintz} & 3 \cdot 3 + 3 \cdot 2 + 2 \cdot 1 = 17 \\ \text{Zukoff} & 3 \cdot 3 + 4 \cdot 2 + 1 \cdot 1 = 18 \end{array}$$

so Zukoff would win the Borda count.

(d)   Zukoff beats Perella 5 to 3, and Zukoff beats Mintz 5 to 3. Therefore, Zukoff is the Condorcet winner.

(e)   The approval vote tally would be

$$\begin{array}{ll} \text{Perella} & 3 \\ \text{Mintz} & 5 \\ \text{Zukoff} & 4 \end{array}$$

so Mintz would win.

(f) Yes, if the two voters who ranked Mintz first and Zukoff second instead ranked Mintz first, Perella second, and Zukoff third, then the Borda counts would be

$$
\begin{array}{ll}
\text{Perella} & 13 + 2 = 15 \\
\text{Mintz} & 17 \\
\text{Zukoff} & 18 - 2 = 16
\end{array}
$$

so Mintz would win.

(g) Yes, if the two voters who ranked Zukoff first and Mintz second and approved of two candidates instead approved of only Zukoff, then the approval vote tally would be

$$
\begin{array}{ll}
\text{Perella} & 3 \\
\text{Mintz} & 3 \\
\text{Zukoff} & 4
\end{array}
$$

so Zukoff would win.

9. Mimicking Condorcet's example with three candidates, an easy example is the following.

Number of Voters

	1	1	1	1
A	1	4	3	2
B	2	1	4	3
C	3	2	1	4
D	4	3	2	1

A would defeat B 3 to 1, B would defeat C 3 to 1, C would defeat D 3 to 1, and D would defeat A 3 to 1.

CHAPTER 2
Student Solution Manual

SECTION 2.1

1. The natural divisor is $23{,}614/45 = 524.7556$.

(a)

Country	Population (in thousands)	Natural Quota	Initial Allocation	Final Allocation
Denmark	5,188	9.8865	9	10
Finland	5,069	9.6597	9	10
Iceland	264	0.5031	0	0
Norway	4,315	8.2229	8	8
Sweden	8,778	16.7278	16	17
Total	23,614		42	45

(b)

Country	Population (in thousands)	Natural Quota	Initial Allocation	Relative Fractional Part	Final Allocation
Denmark	5,188	9.8865	9	0.09850	10
Finland	5,069	9.6597	9	0.07330	10
Iceland	264	0.5031	0	undefined	1
Norway	4,315	8.2229	8	0.02786	8
Sweden	8,778	16.7278	16	0.04549	16
Total	23,614		42		45

3. The natural divisor is $20{,}681/26 \approx 795.4231$.

(a)

Company	New Worth (in thousands)	Natural Quota	Initial Allocation	Final Allocation
Alpha Software	8310	10.4473	10	10
Beta Technology	1073	1.3490	1	1
Gamma Computing	6757	8.4949	8	9
Delta Development	4541	5.7089	5	6
Total	20,681		24	26

(b)

Company	New Worth (in thousands)	Natural Quota	Initial Allocation	Relative Fractional Part	Final Allocation
Alpha Software	8310	10.4473	10	0.04473	10
Beta Technology	1073	1.3490	1	0.34900	2
Gamma Computing	6757	8.4949	8	0.06186	8
Delta Development	4541	5.7089	5	0.14178	6
Total	20,681		24		26

(c) The shift was from a larger company to a smaller company.

5. The natural divisor is $40{,}201/50 \approx 804.0200$.

(a)

District	Number of Eligible Voters	Natural Quota	Initial Allocation	Final Allocation
1	7,478	9.3008	9	9
2	9,003	11.1975	11	11
3	5,397	6.7125	6	7
4	8,825	10.9761	10	11
5	3,562	4.4302	4	5
6	5,936	7.3829	7	7
Total	40,201		47	50

(b)

District	Number of Eligible Voters	Natural Quota	Initial Allocation	Relative Fractional Part	Final Allocation
1	7,478	9.3008	9	0.03342	9
2	9,003	11.1975	11	0.01795	11
3	5,397	6.7125	6	0.11875	7
4	8,825	10.9761	10	0.09761	11
5	3,562	4.4302	4	0.10755	5
6	5,936	7.3829	7	0.05470	7
Total	40,201		47		50

7. The natural divisor is $141{,}885/28 \approx 5067.3214$.

(a)

County	Population	Natural Quota	Initial Allocation	Final Allocation
Cheshire	28,772	5.6780	5	6
Grafton	13,472	2.6586	2	3
Hillsborough	32,871	6.4869	6	6
Rockingham	43,169	8.5191	8	8
Strafford	23,601	4.6575	4	5
Total	141,885		25	28

(b)

County	Population	Natural Quota	Initial Allocation	Relative Fractional Part	Final Allocation
Cheshire	28,772	5.6780	5	0.13560	6
Grafton	13,472	2.6586	2	0.32930	3
Hillsborough	32,871	6.4869	6	0.08115	6
Rockingham	43,169	8.5191	8	0.06489	8
Strafford	23,601	4.6575	4	0.16438	5
Total	141,885		25		28

(c) The three counties with the largest fractional parts were also the smallest three counties.

9. The natural divisor is $94.5/120 \approx 0.7875$.

(a)

Party	Percentage of Votes	Natural Quota	Initial Allocation	Final Allocation
Labor	34.8	44.1905	44	44
Likud	24.9	31.6190	31	32
Energy	9.2	11.6825	11	12
Zionist	5.7	7.2381	7	7
Sephardic Jews	5.1	6.4762	6	7
National Religious	5.0	6.3492	6	6
United Torah	3.4	4.3175	4	4
Democratic Front	2.5	3.1746	3	3
Fatherland	2.3	2.9206	2	3
Arab Democratic	1.6	2.0317	2	2
Total	94.5		116	120

(b)

Party	Percentage of Votes	Natural Quota	Initial Allocation	Relative Fractional Part	Final Allocation
Labor	34.8	44.1905	44	0.00433	44
Likud	24.9	31.6190	31	0.01997	31
Energy	9.2	11.6825	11	0.06205	12
Zionist	5.7	7.2381	7	0.03401	7
Sephardic Jews	5.1	6.4762	6	0.07937	7
National Religious	5.0	6.3492	6	0.05820	6
United Torah	3.4	4.3175	4	0.07938	5
Democratic Front	2.5	3.1746	3	0.05820	3
Fatherland	2.3	2.9206	2	0.46030	3
Arab Democratic	1.6	2.0317	2	0.01585	2
Total	94.5		116		120

(c) no

11. $$\text{state's natural quota} = \frac{\text{state's population}}{\text{natural divisor}} = \frac{\text{state's population}}{\text{total population/house size}}$$

$$= \text{state's population} \cdot \frac{\text{house size}}{\text{total population}}$$

$$= \frac{\text{state's population}}{\text{total population}} \cdot \text{house size}$$

13. Because Hamilton's method rounds state A's quota up and state B's quota down, state A must have a larger fractional part than state B. Because state A is smaller than state B, A's initial allocation will be at most that of B. Thus, the relative fractional part of A's quota will have a larger numerator and a denominator that is equal to or smaller than those of B. It follows that B's initial allocation will round up after A's under Lowndes' method.

SECTION 2.2

1. The natural divisor is $267{,}841/9 \approx 29{,}760.1111$.

    (a)

Candidate	Number of Votes	Natural Quota	Initial Allocation
Grover Cleveland	100,589	3.3800	3
Benjamin Harrison	122,736	4.1242	4
James R. Weaver	30,399	1.0215	1
John Bidwell	14,117	0.4744	0
Total	267,841		8

    The initial allocation is one seat short of the 9 seats we need to allocate. We should expect Cleveland to gain the next seat even though his fractional part is not quite as large as Bidwell's because Cleveland's population is quite a bit larger. Computing both Cleveland's and Bidwell's threshold divisors, we find

    $$\text{Cleveland's threshold divisor} = \frac{100{,}589}{4} = 25{,}147.25,$$

    $$\text{Bidwell's threshold divisor} = \frac{14{,}117}{1} = 14{,}117.$$

    We now use the divisor $D = 25{,}147.25$, and a correct allocation of seats results.

Candidate	Number of Votes	Natural Quota $D = 29{,}760.1111$	Initial Allocation	Modified Quota $D = 25{,}147.25$	Final Allocation
Grover Cleveland	100,589	3.3800	3	4.0000	4
Benjamin Harrison	122,736	4.1242	4	4.8807	4
James R. Weaver	30,399	1.0215	1	1.2088	1
John Bidwell	14,117	0.4744	0	0.5614	0
Total	267,841		8		9

    (b)

Candidate	Number of Votes	Natural Quota	Initial Allocation
Grover Cleveland	100,589	3.3800	3
Benjamin Harrison	122,736	4.1242	4
James R. Weaver	30,399	1.0215	1
John Bidwell	14,117	0.4744	0
Total	267,841		8

    The initial allocation is one seat short of the 9 seats we need to allocate. We compute the threshold divisors for the two candidates with fractional parts that are closest to 0.5.

    $$\text{Cleveland's threshold divisor} = \frac{100{,}589}{3.5} \approx 28{,}739.7143,$$

    $$\text{Bidwell's threshold divisor} = \frac{14{,}117}{0.5} = 28{,}234.$$

    We now use the divisor $D = 28{,}739$ and a correct allocation of seats results.

Candidate	Number of Votes	Natural Quota $D=29{,}760.1111$	Initial Allocation	Modified Quota $D=28{,}739$	Final Allocation
Grover Cleveland	100,589	3.3800	3	3.5001	4
Benjamin Harrison	122,736	4.1242	4	4.2707	4
James R. Weaver	30,399	1.0215	1	1.0578	1
John Bidwell	14,117	0.4744	0	0.4912	0
Total	267,841		8		9

3. The natural divisor is $2213/27 \approx 81.9630$.

(a)

School	Enrollment	Natural Quota	Initial Allocation
Eddyville	60	0.7320	0
Newport	647	7.8938	7
Taft	704	8.5892	8
Toledo	456	5.5635	5
Waldport	346	4.2214	4
Total	2213		24

The initial allocation is three seats less than the 27 seats we need to allocate. We compute the threshold divisors for the four schools with the largest fractional parts.

$$\text{Eddyville's threshold divisor} = \frac{60}{1} = 60,$$

$$\text{Newport's threshold divisor} = \frac{647}{8} = 80.875,$$

$$\text{Taft's threshold divisor} = \frac{704}{9} \approx 78.2222,$$

$$\text{Toledo's threshold divisor} = \frac{456}{6} = 76.$$

Using the divisor $D = 76$, we get the correct allocation of seats.

School	Enrollment	Natural Quota $D=81.9630$	Initial Allocation	Modified Quota $D=69$	Final Allocation
Eddyville	60	0.7320	0	0.7895	0
Newport	647	7.8938	7	8.5132	8
Taft	704	8.5892	8	9.2632	9
Toledo	456	5.5635	5	6.0000	6
Waldport	346	4.2214	4	4.5526	4
Total	2213		24		27

(b)

School	Enrollment	Natural Quota	Initial Allocation
Eddyville	60	0.7320	1
Newport	647	7.8938	8
Taft	704	8.5892	9
Toledo	456	5.5635	6
Waldport	346	4.2214	4
Total	2213		28

The initial allocation is one seat more than the 27 seats we need to allocate. We compute the threshold divisors for the two candidates that appear most likely to lose a seat.

$$\text{Taft's threshold divisor} = \frac{704}{8.5} \approx 82.8235,$$

$$\text{Toledo's threshold divisor} = \frac{456}{5.5} \approx 82.9091.$$

Using $D = 82.85$, a correct allocation of the seats results.

School	Enrollment	Natural Quota $D=81.9630$	Initial Allocation	Modified Quota $D=82.85$	Final Allocation
Eddyville	60	0.7320	1	0.7242	1
Newport	647	7.8938	8	7.8093	8
Taft	704	8.5892	9	8.4973	8
Toledo	456	5.5635	6	5.5039	6
Waldport	346	4.2214	4	4.1762	4
Total	2213		28		27

(c) When switching from Jefferson's method to Webster's method, the shift in seats goes from a school with more students to one with fewer students.

5. The natural divisor is $1{,}108{,}229/42 \approx 26{,}386.4048$.

(a)

County	Population	Natural Quota	Initial Allocation
Hawaii	120,317	4.5598	4
Honolulu	836,231	31.6917	31
Kalawao	130	0.0049	0
Kauai	51,177	1.9395	1
Maui	100,374	3.8040	3
Total	1,108,229		39

The initial allocation is three seats short of the 42 we need to allocate. We compute the threshold divisors of the four counties with the largest fractional parts. However, because Honolulu is so large, we also compute the threshold divisor for Honolulu to gain two more seats because there is a strong possibility that it will gain two seats in this situation.

$$\text{Hawaii's threshold divisor} = \frac{120{,}317}{5} = 24{,}063.4,$$

$$\text{Honolulu's threshold divisor to gain one seat} = \frac{836{,}231}{32} \approx 26{,}132.2188,$$

$$\text{Honolulu's threshold divisor to gain two seats} = \frac{836{,}231}{33} \approx 25{,}340.3333,$$

$$\text{Kauai's threshold divisor} = \frac{51{,}177}{2} = 25{,}588.5,$$

$$\text{Maui's threshold divisor} = \frac{100{,}374}{4} = 25{,}093.5.$$

Using the divisor $D = 25{,}340$, we get a correct allocation. Notice that, as we anticipated could happen, Honolulu did gain two seats, and the Quota Property is violated.

County	Population	Natural Quota $D=26{,}386.4048$	Initial Allocation	Modified Quota $D=25{,}340$	Final Allocation
Hawaii	120,317	4.5598	4	4.7481	4
Honolulu	836,231	31.6917	31	33.0004	33
Kalawao	130	0.0049	0	0.0051	0
Kauai	51,177	1.9395	1	2.0196	2
Maui	100,374	3.8040	3	3.9611	3
Total	1,108,229		39		42

(b)

County	Population	Natural Quota	Initial Allocation
Hawaii	120,317	4.5598	5
Honolulu	836,231	31.6917	32
Kalawao	130	0.0049	0
Kauai	51,177	1.9395	2
Maui	100,374	3.8040	4
Total	1,108,229		43

The initial allocation is one seat over the 42 seats we need to allocate. Because Honolulu has a very large population and its fractional part is not much over 0.5, we should expect it to lose the seat. To be careful in choosing a divisor, we compute the threshold divisors for both Honolulu and Hawaii and find

$$\text{Hawaii's threshold divisor} = \frac{120{,}317}{4.5} \approx 26{,}737.1111,$$

$$\text{Honolulu's threshold divisor} = \frac{836{,}231}{31.5} \approx 26{,}547.0159.$$

Letting $D = 26{,}548$, we get a correct allocation.

County	Population	Natural Quota $D=26{,}386.4048$	Initial Allocation	Modified Quota $D=26{,}548$	Final Allocation
Hawaii	120,317	4.5598	5	4.5321	5
Honolulu	836,231	31.6917	32	31.4988	31
Kalawao	130	0.0049	0	0.0049	0
Kauai	51,177	1.9395	2	1.9277	2
Maui	100,374	3.8040	4	3.7808	4
Total	1,108,229		43		42

(c)

County	Population	Natural Quota	Initial Allocation	Final Allocation
Hawaii	120,317	4.5598	4	4
Honolulu	836,231	31.6917	31	32
Kalawao	130	0.0049	0	0
Kauai	51,177	1.9395	1	2
Maui	100,374	3.8040	3	4
Total	1,108,229		39	42

(d)

County	Population	Natural Quota	Initial Allocation	Relative Fractional Part	Final Allocation
Hawaii	120,317	4.5598	4	0.13995	4
Honolulu	836,231	31.6917	31	0.02231	31
Kalawao	130	0.0049	0	undefined	1
Kauai	51,177	1.9395	1	0.93950	2
Maui	100,374	3.8040	3	0.26800	4
Total	1,108,229		39		42

(e)   Jefferson's method seems to help Honolulu County the most.

(f)   Lowndes' method seems to help Kalawao County the most.

7. The natural divisor is $741/23 \approx 32.2174$.

(a)

Dormitory	Number of Residents	Natural Quota	Initial Allocation
Couch	152	4.7179	4
Galloway	96	2.9798	2
Hardin	147	4.5628	4
Martin	130	4.0351	4
Raney	96	2.9798	2
Veasey	120	3.7247	3
Total	741		19

The initial allocation falls four seats short of the 23 seats we must allocate. We compute the threshold divisors of the five dormitories with the largest fractional parts.

$$\text{Couch's threshold divisor} = \frac{152}{5} = 30.4,$$

$$\text{Galloway's threshold divisor} = \frac{96}{3} = 32,$$

$$\text{Hardin's threshold divisor} = \frac{147}{5} = 29.4,$$

$$\text{Raney's threshold divisor} = \frac{96}{3} = 32,$$

$$\text{Veasey's threshold divisor} = \frac{120}{4} = 30.$$

Letting $D = 30$, we get the correct allocation.

Dormitory	Number of Residents	Natural Quota $D=32.2174$	Initial Allocation	Modified Quota $D=30$	Final Allocation
Couch	152	4.7179	4	5.0667	5
Galloway	96	2.9798	2	3.2000	3
Hardin	147	4.5628	4	4.9000	4
Martin	130	4.0351	4	4.3333	4
Raney	96	2.9798	2	3.2000	3
Veasey	120	3.7247	3	4.0000	4
Total	741		19		23

(b)

Dormitory	Number of Residents	Natural Quota	Initial Allocation
Couch	152	4.7179	5
Galloway	96	2.9798	3
Hardin	147	4.5628	5
Martin	130	4.0351	4
Raney	96	2.9798	3
Veasey	120	3.7247	4
Total	741		24

One too many seats are allocated initially. We expect Hardin, with a large quota that barely rounds up to lose the first seat. To help us in selecting a divisor, we compute the threshold divisors for Hardin and Couch, which seems to be second in line to lose a seat, and find

$$\text{Couch's threshold divisor} = \frac{152}{4.5} \approx 33.7778,$$

$$\text{Hardin's threshold divisor} = \frac{147}{4.5} = 32.6667$$

Letting $D = 33$, we get a correct allocation.

Dormitory	Number of Residents	Natural Quota $D=32.2174$	Initial Allocation	Modified Quota $D=33$	Final Allocation
Couch	152	4.7179	5	4.6061	5
Galloway	96	2.9798	3	2.9091	3
Hardin	147	4.5628	5	4.4545	4
Martin	130	4.0351	4	3.9394	4
Raney	96	2.9798	3	2.9091	3
Veasey	120	3.7247	4	3.6364	4
Total	741		24		23

9. The natural divisor is $94.3/101 \approx 0.9337$.

(a)

Party	Percentage of Votes	Natural Quota	Initial Allocation
Social Democrats	39.8	42.6261	42
Christian Democrats	38.7	41.4480	41
Free Democrats	8.9	9.5320	9
Greens	6.9	7.3900	7
Total	94.3		99

The initial allocation is two seats less than the 101 seats we need to allocate. We compute the threshold divisors for the three parties with largest fractional parts. Note that we should expect the Christian Democrats to receive one of the new seats rather than the Free Democrats because although the Free Democrats have a slightly larger fractional part, the Christian Democrats have a much larger population.

$$\text{Social Democrats' threshold divisor} = \frac{39.8}{43} \approx 0.9256,$$

$$\text{Christian Democrats' threshold divisor} = \frac{38.7}{42} \approx 0.9214,$$

$$\text{Free Democrats' threshold divisor} = \frac{8.9}{10} = 0.89.$$

Letting $D = 0.92$, we arrive at a correct apportionment.

Party	Percentage of Votes	Natural Quota $D = 0.9337$	Initial Allocation	Modified Quota $D = 0.92$	Final Allocation
Social Democrats	39.8	42.6261	42	43.2609	43
Christian Democrats	38.7	41.4480	41	42.0652	42
Free Democrats	8.9	9.5320	9	9.6739	9
Greens	6.9	7.3900	7	7.5000	7
Total	94.3		99		

(b)

Party	Percentage of Votes	Natural Quota	Initial and Final Allocation
Social Democrats	39.8	42.6261	43
Christian Democrats	38.7	41.4480	41
Free Democrats	8.9	9.5320	10
Greens	6.9	7.3900	7
Total	94.3		101

The initial allocation gives a correct apportionment of the 101 seats, so it is also the final allocation.

(c) Neither of the apportionments violate the Quota Property.

(d) Yes, the apportionment by Webster's method agrees with the actual apportionment.

11. The natural divisor is $19{,}374{,}202/400 = 48{,}435.505$.

(a)

Party	Number of Votes	Natural Quota	Initial Allocation
African Nat. Cong.	12,237,655	252.6588	252
National Party	3,983,690	82.2473	82
Inkatha Freedom Party	2,058,294	42.4956	42
Freedom Front	424,555	8.7654	8
Democratic Party	338,426	6.9871	6
Pan Africanist Cong.	243,478	5.0268	5
African Ch. Dem. Party	88,104	1.8190	1
Total	19,374,202		396

The initial allocation falls four seats short of the 400 we need to allocate. We compute the threshold divisors of the five parties with the largest fractional parts. Because they have large populations, we also compute the threshold divisor for the National Party to gain one seat and the threshold divisor for the African National Congress to gain two seats.

$$\text{African National Congress' threshold divisor to gain one seat} = \frac{12{,}237{,}655}{253} \approx 48{,}370.1779,$$

$$\text{African National Congress' threshold divisor to gain two seats} = \frac{12{,}237{,}655}{254} \approx 48{,}179.7441,$$

$$\text{National Party's threshold divisor} = \frac{3{,}983{,}690}{83} \approx 47{,}996.2651,$$

$$\text{Inkatha Freedom Party's threshold divisor} = \frac{2{,}058{,}294}{43} \approx 47{,}867.3023,$$

$$\text{Freedom Front's threshold divisor} = \frac{424{,}555}{9} \approx 47{,}172.7778,$$

$$\text{Democratic Party's threshold divisor} = \frac{338{,}426}{7} \approx 48{,}346.5714,$$

$$\text{African Christian Democratic Party's threshold divisor} = \frac{88{,}104}{2} = 44{,}052.$$

Using $D = 47{,}996$ as our new divisor we get a correct allocation.

Party	Number of Votes	Natural Quota $D = 48{,}435.505$	Initial Allocation	Modified Quota $D = 47{,}996$	Final Allocation
African Nat. Cong.	12,237,655	252.6588	252	254.9724	254
National Party	3,983,690	82.2473	82	83.0005	83
Inkatha Freedom Party	2,058,294	42.4956	42	42.8847	42
Freedom Front	424,555	8.7654	8	8.8456	8
Democratic Party	338,426	6.9871	6	7.0511	7
Pan Africanist Cong.	243,478	5.0268	5	5.0729	5
African Ch. Dem. Party	88,104	1.8190	1	1.8357	1
Total	19,374,202		396		400

(b)

Party	Number of Votes	Natural Quota	Initial and Final Allocation
African Nat. Cong.	12,237,655	252.6588	253
National Party	3,983,690	82.2473	82
Inkatha Freedom Party	2,058,294	42.4956	42
Freedom Front	424,555	8.7654	9
Democratic Party	338,426	6.9871	7
Pan Africanist Cong.	243,478	5.0268	5
African Ch. Dem. Party	88,104	1.8190	2
Total	19,374,202		400

The initial allocation assigns the correct number of seats, so it is also the final allocation.

(c)

Party	Number of Votes	Natural Quota	Initial Allocation	Final Allocation
African Nat. Cong.	12,237,655	252.6588	252	253
National Party	3,983,690	82.2473	82	82
Inkatha Freedom Party	2,058,294	42.4956	42	42
Freedom Front	424,555	8.7654	8	9
Democratic Party	338,426	6.9871	6	7
Pan Africanist Cong.	243,478	5.0268	5	5
African Ch. Dem. Party	88,104	1.8190	1	2
Total	19,374,202		396	400

(d)

Party	Number of Votes	Natural Quota	Initial Allocation	Relative Fractional Part	Final Allocation
African Nat. Cong.	12,237,655	252.6588	252	0.00261	252
National Party	3,983,690	82.2473	82	0.00302	82
Inkatha Freedom Party	2,058,294	42.4956	42	0.01180	43
Freedom Front	424,555	8.7654	8	0.09568	9
Democratic Party	338,426	6.9871	6	0.16452	7
Pan Africanist Cong.	243,478	5.0268	5	0.00536	5
African Ch. Dem. Party	88,104	1.8190	1	0.81900	2
Total	19,374,202		396		400

(e) Yes, the apportionment by Jefferson's method violates the Quota Property.

(f) Yes, the apportionment by Lowndes' method agrees with the actual apportionment.

13. (a)

Region	Number of Barrels (in thousands)	Percent
North America	11,156	17
Central and South America	5,963	9
Western Europe	6,304	10
Eastern Europe and Former U.S.S.R.	6,918	11
Middle East	19,195	30
Africa	7,305	11
Far East and Oceania	7,212	11
Total	64,053	99

(b) The natural divisor is $64,053/100 = 640.53$.

Region	Number of Barrels (in thousands)	Natural Quota	Initial Allocation
North America	11,156	17.4168	17
Central and South America	5,963	9.3095	9
Western Europe	6,304	9.8418	10
Eastern Europe and Former U.S.S.R.	6,918	10.8004	11
Middle East	19,195	29.9674	30
Africa	7,305	11.4046	11
Far East and Oceania	7,212	11.2594	11
Total	64,053		99

The initial allocation is one seat short. We compute the threshold divisors of the two regions that appear closest to gaining a seat.

$$\text{North America's threshold divisor} = \frac{11,156}{17.5} \approx 637.4857,$$

$$\text{Africa's threshold divisor} = \frac{7305}{11.5} \approx 635.2174.$$

Using $D = 637$ results in a correct allocation.

Region	Number of Barrels (in thousands)	Natural Quota $D=640.53$	Initial Allocation	Modified Quota $D=637$	Final Allocation	Percent
North America	11,156	17.4168	17	17.5133	18	18
Central and South America	5,963	9.3095	9	9.3611	9	9
Western Europe	6,304	9.8418	10	9.8964	10	10
Eastern Europe and Former U.S.S.R.	6,918	10.8004	11	10.8603	11	11
Middle East	19,195	29.9674	30	30.1334	30	30
Africa	7,305	11.4046	11	11.4678	11	11
Far East and Oceania	7,212	11.2594	11	11.3218	11	11
Total	64,053		99		100	100

(c)

Region	Number of Barrels (in thousands)	Natural Quota	Initial Allocation
North America	11,156	17.4168	17
Central and South America	5,963	9.3095	9
Western Europe	6,304	9.8418	9
Eastern Europe and Former U.S.S.R.	6,918	10.8004	10
Middle East	19,195	29.9674	29
Africa	7,305	11.4046	11
Far East and Oceania	7,212	11.2594	11
Total	64,053		96

The initial allocation falls four seats short of 100. We compute the threshold divisors for the five regions with largest fractional parts and we also compute the threshold divisor for Middle East to gain two seats because it is large and looks like it will gain a second seat before Africa gains one.

$$\text{North America's threshold divisor} = \frac{11{,}156}{18} \approx 619.7778,$$

$$\text{Western Europe's threshold divisor} = \frac{6304}{10} = 630.4,$$

$$\text{Eastern Europe and Former U.S.S.R.'s threshold divisor} = \frac{6918}{11} \approx 628.9091,$$

$$\text{Middle East's threshold divisor to gain one seat} = \frac{19{,}195}{30} \approx 639.8333,$$

$$\text{Middle East's threshold divisor to gain two seats} = \frac{19{,}195}{31} \approx 619.1935,$$

$$\text{Africa's threshold divisor} = \frac{7305}{12} = 608.75.$$

Using $D = 619.7$, we get a correct allocation.

Region	Number of Barrels (in thousands)	Natural Quota $D = 640.53$	Initial Allocation	Modified Quota $D = 619.7$	Final Allocation	Percent
North America	11,156	17.4168	17	18.0023	18	18
Central and South America	5,963	9.3095	9	9.6224	9	9
Western Europe	6,304	9.8418	9	10.1727	10	10
Eastern Europe and Former U.S.S.R.	6,918	10.8004	10	11.1635	11	11
Middle East	19,195	29.9674	29	30.9747	30	30
Africa	7,305	11.4046	11	11.7880	11	11
Far East and Oceania	7,212	11.2594	11	11.6379	11	11
Total	64,053		96		100	100

15. Because they earned less than 10% of the vote, Alexander, Keyes, Lugar, Gramm, Dornan, and Taylor were not eligible to earn any delegates. Therefore, the 23 delegates must be allocated to Dole, Buchanan, and Forbes. The total population is $52 + 22 + 13 = 87$, and the natural divisor is $87/23 \approx 3.7826$.

(a)

Candidate	Percentage of the Votes	Natural Quota	Initial Allocation
Bob Dole	52	13.7472	13
Pat Buchanan	22	5.8161	5
Steve Forbes	13	3.4368	3
Total	87		21

The initial allocation is two seats short. Computing the threshold divisors of all three candidates we get

$$\text{Dole's threshold divisor} = \frac{52}{14} \approx 3.7143,$$

$$\text{Buchanan's threshold divisor} = \frac{22}{6} \approx 3.6667,$$

$$\text{Forbes' threshold divisor} = \frac{13}{4} = 3.25.$$

Using $D = 3.6$ as our divisor, we get a correct allocation of seats.

Candidate	Percentage of the Votes	Natural Quota $D = 3.7826$	Initial Allocation	Modified Quota $D = 3.6$	Final Allocation
Bob Dole	52	13.7472	13	14.4444	14
Pat Buchanan	22	5.8161	5	6.1111	6
Steve Forbes	13	3.4368	3	3.6111	3
Total	87		21		23

(b)

Candidate	Percentage of the Votes	Natural Quota	Initial and Final Allocation
Bob Dole	52	13.7472	14
Pat Buchanan	22	5.8161	6
Steve Forbes	13	3.4368	3
Total	87		23

The initial allocation assigns the correct number of seats, so it is also the final allocation.

17. The natural divisor is $1{,}108{,}229/42 \approx 26{,}386.4048$.

(a)

County	Population	Natural Quota	Initial Allocation
Hawaii	120,317	4.5598	4
Honolulu	836,231	31.6917	31
Kalawao	130	0.0049	1
Kauai	51,177	1.9395	1
Maui	100,374	3.8040	3
Total	1,108,229		40

We determine the initial allocation as usual except that Kalawao receives one seat even though its natural quota is less than one. The initial allocation is two seats short. We compute the threshold divisors of Kauai, Maui, and Honolulu, and also compute the threshold divisor for the very large county of Honolulu to gain two seats.

$$\text{Honolulu's threshold divisor to gain one seat} = \frac{836{,}231}{32} \approx 26{,}132.2188,$$

$$\text{Honolulu's threshold divisor to gain two seats} = \frac{836{,}231}{33} \approx 25{,}340.3333,$$

$$\text{Kauai's threshold divisor} = \frac{51{,}177}{2} = 25{,}588.5,$$

$$\text{Maui's threshold divisor} = \frac{100{,}374}{4} = 25{,}093.5.$$

Letting $D = 25{,}588$, we get a correct allocation of seats.

County	Population	Natural Quota $D = 26{,}386.4048$	Initial Allocation	Modified Quota $D = 25{,}588$	Final Allocation
Hawaii	120,317	4.5598	4	4.7021	4
Honolulu	836,231	31.6917	31	32.6806	32
Kalawao	130	0.0049	1	0.0051	1
Kauai	51,177	1.9395	1	2.0000	2
Maui	100,374	3.8040	3	3.9227	3
Total	**1,108,229**		40		42

(b)

County	Population	Natural Quota	Initial Allocation
Hawaii	120,317	4.5598	5
Honolulu	836,231	31.6917	32
Kalawao	130	0.0049	1
Kauai	51,177	1.9395	2
Maui	100,374	3.8040	4
Total	**1,108,229**		44

We determine the initial allocation as usual except that Kalawao is allocated a seat even though its natural quota is less than 0.5. The initial allocation is two seats over the 42 seats we need to allocate. We compute the threshold divisors for Hawaii and Honolulu, the two counties that appear most likely to lose a seat, and Maui as well.

$$\text{Hawaii's threshold divisor to lose one seat} = \frac{120{,}317}{4.5} \approx 26{,}737.1111,$$

$$\text{Honolulu's threshold divisor to lose one seat} = \frac{836{,}231}{31.5} \approx 26{,}547.0159,$$

$$\text{Maui's threshold divisor to lose one seat} = \frac{100{,}374}{3.5} \approx 28{,}678.2857.$$

Letting $D = 26{,}738$, we get a correct allocation.

County	Population	Natural Quota $D = 26{,}386.4048$	Initial Allocation	Modified Quota $D = 26{,}738$	Final Allocation
Hawaii	120,317	4.5598	5	4.4999	4
Honolulu	836,231	31.6917	32	31.2750	31
Kalawao	130	0.0049	1	0.0049	1
Kauai	51,177	1.9395	2	1.9140	2
Maui	100,374	3.8040	4	3.7540	4
Total	1,108,229		44		42

19. When using the same divisor, every modified quota that rounds down under Webster's method will also round down under Jefferson's method, as may some others. Thus, with the same divisor, Jefferson's method will allocate at most as many seats as Webster's method, so that a divisor that is too large for Webster's method, allocating too few seats under Webster's method, will also be too large for Jefferson's method.

21. If the natural quotas are integers, all methods we discuss agree. Ignoring this case and the possibility that both fractional parts are exactly 0.5, with two states, one state will have fractional part greater than 0.5 and round up according to Webster's method and the other will have fractional part less than 0.5 and round down. Thus, the natural divisor works for Webster's method with two states. The first state will also gain the additional seat under Hamilton's method.

SECTION 2.3

1. $\sqrt{6 \cdot 7} = \sqrt{42} \approx 6.48074$

3. $\sqrt{780 \cdot 781} = \sqrt{609{,}180} \approx 780.49984$

5. The natural divisor is $1{,}197{,}505/38 \approx 31{,}513.2895$.

Candidate	Number of Votes	Natural Quota	Initial Allocation
Woodrow Wilson	395,637	12.5546	13
Theodore Roosevelt	444,894	14.1177	14
William H. Taft	273,360	8.6744	9
Eugene V. Debs	83,614	2.6533	3
Total			39

None of the natural quotas have fractional parts between 0.4 and 0.5, so we can round without computing any Hill-Huntington cutoffs. The resulting initial allocation is one seat over the thirty-eight seats we need to allocate. Woodrow Wilson will certainly be the candidate to lose a seat because the fractional part of his natural quota is the closest to 0.5, and the only candidate with a larger population, Theodore Roosevelt, has only a slightly larger population and is not close to losing a seat. Computing Wilson's threshold divisor and that of Taft who appears next most likely to lose a seat we find

$$\text{Wilson's threshold divisor} = \frac{395{,}637}{\sqrt{12 \cdot 13}} = \frac{395{,}637}{\sqrt{156}} \approx 31{,}676.3112,$$

$$\text{Taft's threshold divisor} = \frac{273{,}360}{\sqrt{8 \cdot 9}} = \frac{273{,}360}{\sqrt{72}} \approx 32{,}215.7850.$$

Letting $D = 31{,}677$ and using the fact that the cutoff between 12 and 13 seats is $\sqrt{12 \cdot 13} = \sqrt{156} \approx 12.49000$, we see that the allocation is correct.

Candidate	Number of Votes	Natural Quota $D=31{,}513.2895$	Initial Allocation	Modified Quota $D=31{,}677$	Final Allocation
Woodrow Wilson	395,637	12.5546	13	12.4897	12
Theodore Roosevelt	444,894	14.1177	14	14.0447	14
William H. Taft	273,360	8.6744	9	8.6296	9
Eugene V. Debs	83,614	2.6533	3	2.6396	3
Total	1,197,505		39		38

7. The natural divisor is $3{,}287{,}116/79 \approx 41{,}609.0633$.

County	Population	Natural Quota	Initial Allocation
Fairfield	827,645	19.8910	20
Hartford	851,783	20.4711	20
Litchfield	174,092	4.1840	4
Middlesex	143,196	3.4415	3
New Haven	804,219	19.3280	19
New London	254,957	6.1274	6
Tolland	128,699	3.0931	3
Windham	102,525	2.4640	3
Total	3,287,116		78

The cutoff between 20 and 21 seats is $\sqrt{20 \cdot 21} = \sqrt{420} \approx 20.49390$, the cutoff between 3 and 4 seats is $\sqrt{3 \cdot 4} = \sqrt{12} \approx 3.46410$, and the cutoff between 2 and 3 seats is $\sqrt{2 \cdot 3} = \sqrt{6} \approx 2.44949$. Allocating the seats, we see that the initial allocation falls one seat short. We compute the threshold divisors of the two counties that appear closest to gaining a seat.

$$\text{Hartford's threshold divisor} = \frac{851{,}783}{\sqrt{20 \cdot 21}} = \frac{851{,}783}{\sqrt{420}} \approx 41{,}562.75459,$$

$$\text{Middlesex's threshold divisor} = \frac{143{,}196}{\sqrt{4 \cdot 5}} = \frac{143{,}196}{\sqrt{20}} \approx 32{,}019.59901.$$

Letting $D = 41{,}562$, we have a correct allocation of seats.

County	Population	Natural Quota $D=41{,}609.0633$	Initial Allocation	Modified Quota $D=41{,}562$	Final Allocation
Fairfield	827,645	19.8910	20	19.9135	20
Hartford	851,783	20.4711	20	20.4943	21
Litchfield	174,092	4.1840	4	4.1887	4
Middlesex	143,196	3.4415	3	3.4454	3
New Haven	804,219	19.3280	19	19.3499	19
New London	254,957	6.1274	6	6.1344	6
Tolland	128,699	3.0931	3	3.0966	3
Windham	102,525	2.4640	3	2.4668	3
Total	3,287,116		78		79

9. The natural divisor is $147{,}612/44 \approx 3354.8182$.

(a)

County	Population (in thousands)	Natural Quota	Initial Allocation
Algeria	31,743	9.4619	9
Egypt	67,542	20.1328	20
Libya	6,294	1.8761	2
Morocco	32,189	9.5949	10
Tunisia	9,599	2.8613	3
Western Sahara	245	0.0730	1
Total	147,612		45

The cutoff between 9 and 10 seats is $\sqrt{9 \cdot 10} = \sqrt{90} \approx 9.48683$. Allocating the seats, we see that the initial allocation assigns one seat too many. Morocco will almost certainly lose the seat. Computing its threshold divisor and that of Egypt which has a large population, we find

$$\text{Egypt's threshold divisor} = \frac{67{,}542}{\sqrt{19 \cdot 20}} = \frac{67{,}542}{\sqrt{380}} \approx 3464.8315,$$

$$\text{Morocco's threshold divisor} = \frac{32{,}189}{\sqrt{9 \cdot 10}} = \frac{32{,}189}{\sqrt{90}} \approx 3393.0185.$$

Letting $D = 3394$ and again using the fact that the cutoff between 9 and 10 seats is 9.48683, we get a correct allocation of seats.

Country	Population (in thousands)	Natural Quota $D = 3354.8182$	Initial Allocation	Modified Quota $D = 3394$	Final Allocation
Algeria	31,743	9.4619	9	9.3527	9
Egypt	67,542	20.1328	20	19.9004	20
Libya	6,294	1.8761	2	1.8544	2
Morocco	32,189	9.5949	10	9.4841	9
Tunisia	9,599	2.8613	3	2.8282	3
Western Sahara	245	0.0730	1	0.0722	1
Total	147,612		45		44

(b)

County	Population (in thousands)	Natural Quota	Initial Allocation	Final Allocation
Algeria	31,743	9.4619	9	9
Egypt	67,542	20.1328	20	20
Libya	6,294	1.8761	1	2
Morocco	32,189	9.5949	9	10
Tunisia	9,599	2.8613	2	3
Western Sahara	245	0.0730	0	0
Total	147,612		41	44

(c)

County	Population (in thousands)	Natural Quota	Initial Allocation	Relative Fractional Part	Final Allocation
Algeria	31,743	9.4619	9	0.05132	9
Egypt	67,542	20.1328	20	0.00664	20
Libya	6,294	1.8761	1	0.87610	2
Morocco	32,189	9.5949	9	0.06610	9
Tunisia	9,599	2.8613	2	0.43065	3
Western Sahara	245	0.0730	0	undefined	1
Total	147,612		41		44

(d)

County	Population (in thousands)	Natural Quota	Initial Allocation
Algeria	31,743	9.4619	9
Egypt	67,542	20.1328	20
Libya	6,294	1.8761	1
Morocco	32,189	9.5949	9
Tunisia	9,599	2.8613	2
Western Sahara	245	0.0730	0
Total	147,612		41

The initial allocation falls three seats short. Because some the states with the largest fractional parts are not the largest states, it is not clear which countries will gain seats. Therefore, we compute the threshold divisors for all the countries except Western Sahara.

$$\text{Algeria's threshold divisor} = \frac{31,743}{10} = 3174.3,$$

$$\text{Egypt's threshold divisor} = \frac{67,542}{21} \approx 3216.2857,$$

$$\text{Libya's threshold divisor} = \frac{6294}{2} = 3147,$$

$$\text{Morocco's threshold divisor} = \frac{32,189}{10} = 3218.9,$$

$$\text{Tunisia's threshold divisor} = \frac{9599}{3} \approx 3199.6667.$$

Using $D = 3199$, we get a correct allocation of seats.

Country	Population (in thousands)	Natural Quota $D = 3354.8182$	Initial Allocation	Modified Quota $D = 3394$	Final Allocation
Algeria	31,743	9.4619	9	9.9228	9
Egypt	67,542	20.1328	20	21.1135	21
Libya	6,294	1.8761	1	1.9675	1
Morocco	32,189	9.5949	9	10.0622	10
Tunisia	9,599	2.8613	2	3.0006	3
Western Sahara	245	0.0730	0	0.0766	0
Total	147,612		41		44

(e)

County	Population (in thousands)	Natural Quota	Initial and Final Allocation
Algeria	31,743	9.4619	9
Egypt	67,542	20.1328	20
Libya	6,294	1.8761	2
Morocco	32,189	9.5949	10
Tunisia	9,599	2.8613	3
Western Sahara	245	0.0730	0
Total	147,612		44

All 44 seats are allocated initially, so we have a correct allocation of seats.

(f) Jefferson's method seems to help Egypt the most.

11. The natural divisor is $9026/175 \approx 51.5771$.

(a)

Precinct	Number of Crimes	Natural Quota	Initial & Final Allocation
1	3015	58.4562	58
2	624	12.0984	12
3	1775	34.4145	34
4	1479	28.6755	29
5	1212	23.4988	24
6	921	17.8568	18
Total	9026		175

The cutoff between 58 and 59 seats is $\sqrt{58 \cdot 59} = \sqrt{3422} \approx 58.49786$, the cutoff between 34 and 35 seats is $\sqrt{34 \cdot 35} = \sqrt{1190} \approx 34.49638$, and the cutoff between 23 and 24 seats is $\sqrt{23 \cdot 24} = \sqrt{552} \approx 23.49468$. Allocating the seats, we see that the initial allocation assigns the correct number of seats, so it is also the final allocation.

(b)

Precinct	Number of Crimes	Natural Quota	Initial Allocation	Final Allocation
1	3015	58.4562	58	58
2	624	12.0984	12	12
3	1775	34.4145	34	34
4	1479	28.6755	28	29
5	1212	23.4988	23	24
6	921	17.8568	17	18
Total	9026		172	175

(c)

Precinct	Number of Crimes	Natural Quota	Initial Allocation	Relative Fractional Part	Final Allocation
1	3015	58.4562	58	0.00787	58
2	624	12.0984	12	0.00820	12
3	1775	34.4145	34	0.01219	34
4	1479	28.6755	28	0.02413	29
5	1212	23.4988	23	0.02169	24
6	921	17.8568	17	0.05040	18
Total	9026		172		175

(d)

Precinct	Number of Crimes	Natural Quota	Initial Allocation
1	3015	58.4562	58
2	624	12.0984	12
3	1775	34.4145	34
4	1479	28.6755	28
5	1212	23.4988	23
6	921	17.8568	17
Total	9026		172

The initial allocation is three seats short of the 175 we need to allocate. Precinct 2 is unlikely to gain a seat. Computing the threshold divisors of the other five precincts, we find that

$$\text{Precinct 1's threshold divisor} = \frac{3015}{59} \approx 51.1017,$$

$$\text{Precinct 3's threshold divisor} = \frac{1775}{35} = 50.7143,$$

$$\text{Precinct 4's threshold divisor} = \frac{1479}{29} = 51,$$

$$\text{Precinct 5's threshold divisor} = \frac{1212}{24} = 50.5,$$

$$\text{Precinct 6's threshold divisor} = \frac{921}{18} \approx 51.1667.$$

Letting $D = 51$, we get a correct allocation of seats.

Precinct	Number of Crimes	Natural Quota $D = 51.5771$	Initial Allocation	Modified Quota $D = 51$	Final Allocation
1	3015	58.4562	58	59.1176	59
2	624	12.0984	12	12.2353	12
3	1775	34.4145	34	34.8039	34
4	1479	28.6755	28	29.0000	29
5	1212	23.4988	23	23.7647	23
6	921	17.8568	17	18.0588	18
Total	9026		172		175

(e)

Precinct	Number of Crimes	Natural Quota	Initial Allocation
1	3015	58.4562	58
2	624	12.0984	12
3	1775	34.4145	34
4	1479	28.6755	29
5	1212	23.4988	23
6	921	17.8568	18
Total	9026		174

The initial allocation leaves one more seat to be assigned. The seat will certainly go to either Precinct 1 or Precinct 5, so we compute their threshold divisors and get

$$\text{Precinct 1's threshold divisor} = \frac{3015}{58.5} \approx 51.5385,$$

$$\text{Precinct 5's threshold divisor} = \frac{1212}{23.5} \approx 51.5745.$$

Using the divisor $D = 51.57$ gives a correct allocation of seats.

Precinct	Number of Crimes	Natural Quota $D = 51.5771$	Initial Allocation	Modified Quota $D = 51.57$	Final Allocation
1	3015	58.4562	58	58.4642	58
2	624	12.0984	12	12.1001	12
3	1775	34.4145	34	34.4192	34
4	1479	28.6755	29	28.6795	29
5	1212	23.4988	23	23.5020	24
6	921	17.8568	18	17.8592	18
Total	9026		174		175

13.   The natural divisor is $95.9/99 \approx 0.9687$.

Party	Percentage of Votes	Natural Quota	Initial Allocation
Red Party	31.4	32.4146	32
National White	30.2	31.1758	31
Prog. Encounter	29.3	30.2467	30
New Space	5.0	5.1616	5
Total	95.9		98

The cutoff between 32 and 33 seats is $\sqrt{32 \cdot 33} = \sqrt{1056} \approx 32.49615$. Allocating the seats, we see that the initial allocation is one seat short. This seat will certainly go to the Red Party because it has a fractional part closest to the cutoff for a new seat and it is the largest party. Computing its threshold divisor and that of the Progressive Encounter Party which appears next in line to gain a seat, we find

$$\text{Red Party's threshold divisor} = \frac{31.4}{\sqrt{32 \cdot 33}} = \frac{31.4}{\sqrt{1056}} \approx 0.9663,$$

$$\text{Progressive Encounter's threshold divisor} = \frac{29.3}{\sqrt{30 \cdot 31}} = \frac{29.3}{\sqrt{930}} \approx 0.9608.$$

Letting $D = 0.966$, we get a correct allocation of seats.

Party	Percentage of Votes	Natural Quota $D=0.9687$	Initial Allocation	Modified Quota $D=0.966$	Final Allocation
Red Party	31.4	32.4146	32	32.5052	33
National White	30.2	31.1758	31	31.2629	31
Prog. Encounter	29.3	30.2467	30	30.3313	30
New Space	5.0	5.1616	5	5.1760	5
Total	95.9				99

15. The natural divisor is $3,893,625/105 \approx 37,082.1429$.

State	Population	Natural Quota	Initial Allocation
Connecticut	237,946	6.4167	6
Delaware	59,094	1.5936	2
Georgia	82,538	2.2258	2
Kentucky	73,677	1.9869	2
Maine	96,540	2.6034	3
Maryland	319,728	8.6222	9
Massachusetts	378,787	10.2148	10
New Hampshire	141,885	3.8262	4
New Jersey	184,139	4.9657	5
New York	340,120	9.1721	9
North Carolina	393,751	10.6183	11
Pennsylvania	434,373	11.7138	12
Rhode Island	68,825	1.8560	2
South Carolina	249,073	6.7168	7
Vermont	85,539	2.3067	2
Virginia	747,610	20.1609	20
Total	3,893,625		106

The cutoff between 6 and 7 seats is $\sqrt{6 \cdot 7} = \sqrt{42} \approx 6.48074$. Allocating the seats, we see that the initial allocation assigns one seat too many. North Carolina looks likely to lose a seat because it has a fractional part fairly close to the cutoff and it has a large population. Computing its threshold divisor and that of Maryland which appears next closest to losing a seat, we get

$$\text{Maryland's threshold divisor} = \frac{319,728}{\sqrt{8 \cdot 9}} = \frac{319,728}{\sqrt{72}} \approx 37,680.3062,$$

$$\text{North Carolina's threshold divisor} = \frac{393,751}{\sqrt{10 \cdot 11}} = \frac{393,751}{\sqrt{110}} \approx 37,542.6848.$$

Letting $D = 37,543$ and using the fact that the cutoff between 10 and 11 seats is $\sqrt{10 \cdot 11} = \sqrt{110} \approx 10.48809$, we get a correct allocation of seats.

State	Population	Natural Quota $D = 37{,}082.1429$	Initial Allocation	Modified Quota $D = 37{,}543$	Final Allocation
Connecticut	237,946	6.4167	6	6.3380	6
Delaware	59,094	1.5936	2	1.5740	2
Georgia	82,538	2.2258	2	2.1985	2
Kentucky	73,677	1.9869	2	1.9625	2
Maine	96,540	2.6034	3	2.5715	3
Maryland	319,728	8.6222	9	8.5163	9
Massachusetts	378,787	10.2148	10	10.0894	10
New Hampshire	141,885	3.8262	4	3.7793	4
New Jersey	184,139	4.9657	5	4.9047	5
New York	340,120	9.1721	9	9.0595	9
North Carolina	393,751	10.6183	11	10.4880	10
Pennsylvania	434,373	11.7138	12	11.5700	12
Rhode Island	68,825	1.8560	2	1.8332	2
South Carolina	249,073	6.7168	7	6.6343	7
Vermont	85,539	2.3067	2	2.2784	2
Virginia	747,610	20.1609	20	19.9134	20
Total	**3,893,625**		**106**		**105**

17. (a)

Commodity	Millions of Dollars	Percent
Food and agriculture	44,694	7.6
Beverages and tobacco	8,074	1.4
Crude materials	34,820	6.0
Mineral fuels and related products	10,478	1.8
Chemicals and related products	61,755	10.6
Machinery and transport equip.	282,861	48.4
Manufactured goods and other	142,060	24.3
Total	**584,742**	**100.1**

The total of the rounded percentages is 100.1%.

(b) The natural divisor is $584{,}742/1000 = 584.742$.

Commodity	Millions of Dollars	Natural Quota	Initial Allocation
Food and agriculture	44,694	76.4337	76
Beverages and tobacco	8,074	13.8078	14
Crude materials	34,820	59.5476	60
Mineral fuels and related products	10,478	17.9190	18
Chemicals and related products	61,755	105.6107	106
Machinery and transport equip.	282,861	483.7364	484
Manufactured goods and other	142,060	242.9448	243
Total	**584,742**		**1001**

The cutoff between 76 and 77 seats is $\sqrt{76 \cdot 77} = \sqrt{5852} \approx 76.49834$. Allocating the seats, we see that the initial allocation assigns one seat too many. It is not clear which commodity will lose a seat, so we compute the threshold divisors of the three whose fractional parts are closest to the cutoffs and get

$$\text{Crude material's threshold divisor} = \frac{34{,}820}{\sqrt{59 \cdot 60}} = \frac{34{,}820}{\sqrt{3540}} \approx 585.2307,$$

$$\text{Chemicals and related products' threshold divisor} = \frac{61{,}755}{\sqrt{105 \cdot 106}} = \frac{61{,}755}{\sqrt{11{,}130}} \approx 585.3620$$

$$\text{Machinery and transport equipment's threshold divisor} = \frac{282{,}861}{\sqrt{483 \cdot 484}} = \frac{282{,}861}{\sqrt{233{,}772}} \approx 585.0282.$$

Letting $D = 585.1$ and using the fact that the cutoff between 483 and 484 seats is $\sqrt{483 \cdot 484} = \sqrt{233{,}772} \approx 483.49974$, we get a correct allocation of seats.

Commodity	Millions of Dollars	Natural Quota $D = 584.742$	Initial Allocation	Modified Quota $D = 585.1$	Final Allocation	Percent
Food and agriculture	44,694	76.4337	76	76.3869	76	7.6
Beverages and tobacco	8,074	13.8078	14	13.7994	14	1.4
Crude materials	34,820	59.5476	60	59.5112	60	6.0
Mineral fuels and related products	10,478	17.9190	18	17.9080	18	1.8
Chemicals and related products	61,755	105.6107	106	105.5461	106	10.6
Machinery and transport equip.	282,861	483.7364	484	483.4404	483	48.3
Manufactured goods and other	142,060	242.9448	243	242.7961	243	24.3
Total	584,742		1001		1000	100.0

19. The natural divisor is $3{,}287{,}116/79 \approx 41{,}609.0633$.

County	Population	Natural Quota	Initial Allocation
Fairfield	827,645	19.8910	20
Hartford	851,783	20.4711	21
Litchfield	174,092	4.1840	5
Middlesex	143,196	3.4415	4
New Haven	804,219	19.3280	20
New London	254,957	6.1274	7
Tolland	128,699	3.0931	4
Windham	102,525	2.4640	3
Total	3,287,116		84

The initial allocation assigns five seats too many. We compute the threshold divisors of the three counties with the smallest fractional parts and the three counties with the largest populations because it appears that the five seats will be lost by some of these counties.

$$\text{Fairfield's threshold divisor} = \frac{827,645}{19} \approx 43,560.2632,$$

$$\text{Hartford's threshold divisor} = \frac{851,783}{20} = 42,589.15,$$

$$\text{Litchfield's threshold divisor} = \frac{174,092}{4} = 43,523,$$

$$\text{New Haven's threshold divisor} = \frac{804,219}{19} \approx 42,327.3158,$$

$$\text{New London's threshold divisor} = \frac{254,957}{6} \approx 42,492.8333,$$

$$\text{Tolland's threshold divisor} = \frac{128,699}{3} \approx 42,899.6667.$$

Using $D = 43,523$, we get a correct allocation of seats.

County	Population	Natural Quota $D = 41,609.0633$	Initial Allocation	Modified Quota $D = 43,523$	Final Allocation
Fairfield	827,645	19.8910	20	19.0163	20
Hartford	851,783	20.4711	21	19.5709	20
Litchfield	174,092	4.1840	5	4.0000	4
Middlesex	143,196	3.4415	4	3.2901	4
New Haven	804,219	19.3280	20	18.4780	19
New London	254,957	6.1274	7	5.8580	6
Tolland	128,699	3.0931	4	2.9570	3
Windham	102,525	2.4640	3	2.3557	3
Total	3,287,116		84		79

21. threshold divisor for n seats under Dean's method $= \dfrac{\text{population of the state}}{(n-1)n/(n-0.5)}$

23. The natural divisor is $3,287,116/79 \approx 41,609.0633$.

County	Population	Natural Quota	Initial & Final Allocation
Fairfield	827,645	19.8910	20
Hartford	851,783	20.4711	20
Litchfield	174,092	4.1840	4
Middlesex	143,196	3.4415	4
New Haven	804,219	19.3280	19
New London	254,957	6.1274	6
Tolland	128,699	3.0931	3
Windham	102,525	2.4640	3
Total	3,287,116		79

The cutoff between 19 and 20 seats is $19 \cdot 20/19.5 \approx 19.48718$, between 20 and 21 seats it is $20 \cdot 21/20.5 \approx 20.48780$, between 4 and 5 seats it is $4 \cdot 5/4.5 \approx 4.44444$, between 3 and 4 seats it is $3 \cdot 4/3.5 \approx 3.42857$, between 6 and 7 seats it is $6 \cdot 7/6.5 \approx 6.46154$, and between 2 and 3 seats it is $2 \cdot 3/2.5 = 2.4$. Using these cutoffs, we see that exactly 79 are allocated, so the initial allocation is also the final allocation.

25. The natural divisor is $3,287,116/79 \approx 41,609.0633$.

County	Population	Natural Quota	Initial Allocation
Fairfield	827,645	19.8910	20
Hartford	851,783	20.4711	21
Litchfield	174,092	4.1840	4
Middlesex	143,196	3.4415	4
New Haven	804,219	19.3280	19
New London	254,957	6.1274	6
Tolland	128,699	3.0931	3
Windham	102,525	2.4640	3
Total	**3,287,116**		**80**

The initial allocation is one seat over the 79 seats we need to assign. Because Middlesex has fractional part that went over 0.4 by the least amount and because Hartford has fractional part not far above 0.4 and a larger population, we should expect one of these counties to lose the seat. We compute their threshold divisors.

$$\text{Hartford's threshold divisor} = \frac{851,783}{21 - 0.6} = \frac{851,783}{20.4} \approx 41,754.0686,$$

$$\text{Middlesex's threshold divisor} = \frac{143,196}{4 - 0.6} = \frac{143,196}{3.4} \approx 42,116.4706.$$

Using $D = 41,755$ gives a correct allocation of seats.

County	Population	Natural Quota $D=41,609.0633$	Initial Allocation	Modified Quota $D=41,755$	Final Allocation
Fairfield	827,645	19.8910	20	19.8215	20
Hartford	851,783	20.4711	21	20.3995	20
Litchfield	174,092	4.1840	4	4.1694	4
Middlesex	143,196	3.4415	4	3.4294	4
New Haven	804,219	19.3280	19	19.2604	19
New London	254,957	6.1274	6	6.1060	6
Tolland	128,699	3.0931	3	3.0822	3
Windham	102,525	2.4640	3	2.4554	3
Total	**3,287,116**		**80**		**79**

27. Because the Hill-Huntington method rounds at lower values than does Webster's method, with the same divisor, the Hill-Huntington method will allocate at least as many seats as Webster's method. Thus, a divisor that is too small for Webster's method, allocating too many seats under Webster's method, will also be too small for the Hill-Huntington method.

29. Note that the natural divisor works for Webster's method with two states (see Exercise 21 of Section 2.2) and, with the same divisor, the Hill-Huntington method will allocate at least as many seats to each state as Webster's method. For the apportionments to differ, the natural divisor must be too small to work with the Hill-Huntington method. In order for the state whose fractional part is greater than 0.5 to lose the extra seat under the Hill-Huntington method, it should be the larger state and have a fractional part just greater than 0.5. We allocate 10 seats between two states with total population 1000. The Hill-Huntington cutoffs for 2 and 9 seats are $\sqrt{1 \cdot 2} \approx 1.41421$ and $\sqrt{8 \cdot 9} \approx 8.48528$, respectively.

State	Population	Natural Quota $D=100$	Initial & Final Allocation by Webster's Method	Initial Allocation by the Hill-Huntington Method	Modified Quota for the Hill-Huntington Method $D=101$	Final Allocation by the Hill-Huntington Method
A	851	8.51	9	9	8.4257	8
B	149	1.49	1	2	1.4752	2

SECTION 2.4

1.

		Quota Property	House Size Property	Population Property
Quota	Hamilton	✓		
Methods	Lowndes	✓		

Divisor	Jefferson		✓	✓
Methods	Webster		✓	✓
	Hill-Huntington		✓	✓

3. No method satisfies both the Quota Property and the Population Property because only divisor methods satisfy the Population Property and such methods never satisfy the Quota Property.

5. The total population is 50,059 and the natural divisor is $50,059/50 = 1001.18$.

State	Population	Natural Quota	Initial Allocation
A	14,978	14.9603	14
B	12,991	12.9757	12
C	9,260	9.2491	9
D	5,453	5.4466	5
E	4,624	4.6186	4
F	2,753	2.7498	2
Total	50,059		46

The initial allocation using Jefferson's method is 4 seats short of the 50 seats we need to allocate. It is clear that states A and B will gain at least one seat each. Moreover, the wording of the exercise leads us to expect that each will gain a second seat and therefore violate the Quota Property. We compute the threshold divisors for A and B to gain second seats

$$\text{A's threshold divisor for 16 seats} = \frac{14,978}{16} = 936.125,$$

$$\text{B's threshold divisor for 14 seats} = \frac{12,991}{14} \approx 927.9286.$$

We use the divisor 927 and find a final allocation in which states A and B both violate quota.

State	Population	Natural Quota	Initial Allocation	Modified Quota $D=927$	Final Allocation
A	14,978	14.9603	14	16.1575	16
B	12,991	12.9757	12	14.0140	14
C	9,260	9.2491	9	9.9892	9
D	5,453	5.4466	5	5.8824	5
E	4,624	4.6186	4	4.9881	4
F	2,753	2.7498	2	2.9698	2
Total	50,059		46		50

7. With 24 seats, the natural divisor is $5444/24 \approx 226.8333$, and states D and E gain the two seats remaining after the initial allocation.

Apportionment of 24 Seats

State	Population	Natural Quota	Initial Allocation	Final Allocation
A	1676	7.3887	7	7
B	1454	6.4100	6	6
C	921	4.0603	4	4
D	778	3.4298	3	4
E	615	2.7112	2	3
Total	5444		22	24

With 25 seats, the natural divisor is $5444/25 \approx 217.76$, and the initial allocation is three seats short. They go to A, B, and E.

Apportionment of 24 Seats

State	Population	Natural Quota	Initial Allocation	Final Allocation
A	1676	7.6965	7	8
B	1454	6.6771	6	7
C	921	4.2294	4	4
D	778	3.5727	3	3
E	615	2.8242	2	3
Total	5444		22	25

State D gets 4 seats in a 24 seat house, but only 3 in a 25 seat house, a violation of the House Size Property.

9. In the apportionment using Lowndes' method with the old population, the natural divisor is $23,391/23 = 1017$. States C and D have the largest relative fractional parts, hence gain the two seats remaining after the initial allocation.

Apportionment for Old Population

State	Population	Natural Quota	Initial Allocation	Relative Fractional Part	Allocation
A	11,167	10.9803	10	0.09803	10
B	7,536	7.4100	7	0.05857	7
C	3,356	3.2999	3	0.09997	4
D	1,332	1.3097	1	0.30970	2
Total	23,391		21		23

In the initial apportionment using Lowndes' method with the new populations, the natural divisor is
$23,256/23 = 1011.1304$. After the initial allocation, state D gets the one remaining seat.

Apportionment for New Population

State	Population	Natural Quota	Initial Allocation	Relative Fractional Part	Allocation
A	11,160	11.0372	11	0.00338	11
B	7,536	7.4530	7	0.06471	7
C	3,360	3.3230	3	0.10767	3
D	1,200	1.1868	1	0.18680	2
Total	23,256		22		23

State A has lost population and gained a seat, while state C has gained population and lost a seat, a
violation of the Population Property.

11. We construct a large state whose natural quota is just less than an integer, so that it will be a strong
candidate to gain two seats and thus violate the Quota Property. Our house size will be 41.

State	Population	Natural Quota $D = 13.2195$	Initial Allocation	Modified Quota $D = 12.8$	Final Allocation
A	489	36.9908	36	38.2031	38
B	32	2.4207	2	2.5000	2
C	21	1.5886	1	1.6406	1

We see that state A's allocation leads to a violation of the Quota Property.

13. If the natural quotas of the two states are integers, both will be allocated their exact quotas. Otherwise
the (unrounded) fractional parts of the two states will add up to 1. Ignoring the possibility that both
are 0.5, one state's natural quota will round up and one will round down. Thus, the natural quota will
give the correct apportionment by Webster's method and results in an apportionment that satisfies
the Quota Property.

15. When there are two states, Hamilton's method rounds natural quotas in their natural way, up at 0.5,
down below 0.5. When the house size increases, natural quotas also increase, so that neither state's
allocation can decrease.

17. We again want a close call with relative fractional parts, with a seat going to a smaller state. When
the house size increases by 1, we want two larger states to automatically gain seats by having the
integer parts of their natural quotas increase by one. Our house sizes will be 21 and 22.

Apportionment for House Size 21

State	Population	Natural Quota $D = 100.3333$	Relative Fractional Part	Allocation
A	886	8.8306	0.10383	8
B	775	7.7243	0.10347	7
C	112	1.1163	0.11630	2
D	334	3.3289	0.10963	4

Apportionment for House Size 22

State	Population	Natural Quota $D=95.7727$	Relative Fractional Part	Allocation
A	886	9.2511	0.02790	9
B	775	8.0921	0.01151	8
C	112	1.1694	0.16940	2
D	334	3.4874	0.16247	3

Observe that state D loses a seat as the house size increases from 21 to 22.

19. When the total population does not change, the natural quota of a state gaining population increases whereas that of a state losing population decreases. Suppose that this first state loses a seat while the second gains a seat. For the first state's allocation to decrease, the integer part of its natural quota cannot change, and the natural quota must round up at first and down after it gains population. Similarly the integer part of the natural quota of the state losing population cannot change, and its natural quota must round down at first and up after it loses population. However, together these imply that the first state's fractional part is larger than the second's at first but then is smaller after it gains population and the second state loses population. This is impossible.

21. As with most such examples, it is easiest to come up with one with a great variance in the size of the states. Notice that we change the populations of the two relevant states as little as possible. Our house size is 30.

State	Population	Natural Quota $D=92.8$	Relative Fractional Part	Allocation
A	1950	21.0129	0.00061	21
B	730	7.8664	0.12377	8
C	104	1.1207	0.12070	1

State	Population	Natural Quota $D=95.1333$	Relative Fractional Part	Allocation
A	1951	20.5081	0.02541	20
B	800	8.4093	0.05116	8
C	103	1.0827	0.08270	2

We see that state A's population has gone up, yet it has lost a seat, whereas state C's has gone down, yet it has gained a seat, a violation of the Population Property.

Chapter 2 Review Exercises

1. The natural divisor is $534,650/35 \approx 15,275.7143$.

(a)

Company	Number of Patients	Natural Quota	Initial Allocation	Final Allocation
East Michigan	85,750	5.6135	5	6
Great Lakes	142,450	9.3253	9	9
Mid Michigan	68,170	4.4626	4	4
Southeast Michigan	238,280	15.5986	15	16
Total	534,650		33	35

(b)

Company	Number of Patients	Natural Quota	Initial Allocation	Relative Fractional Part	Final Allocation
East Michigan	85,750	5.6135	5	0.12270	6
Great Lakes	142,450	9.3253	9	0.03614	9
Mid Michigan	68,170	4.4626	4	0.11565	5
Southeast Michigan	238,280	15.5986	15	0.03991	15
Total	534,650		33		35

(c)

Company	Number of Patients	Natural Quota	Initial Allocation
East Michigan	85,750	5.6135	5
Great Lakes	142,450	9.3253	9
Mid Michigan	68,170	4.4626	4
Southeast Michigan	238,280	15.5986	15
Total	534,650		33

The initial allocation is two seats short. Because it has a large fractional part and a large population, it appears that Southeast Michigan will gain a seat. The remaining seat will go to East Michigan which has the largest fractional part or to Great Lakes which has a smaller fractional part than East Michigan, but a larger population. We compute the threshold divisors for these three companies.

$$\text{East Michigan's threshold divisor} = \frac{85,750}{6} \approx 14,291.6667,$$

$$\text{Great Lake's threshold divisor} = \frac{142,450}{10} = 14,245,$$

$$\text{Southeast Michigan's threshold divisor} = \frac{238,280}{16} = 14,892.5.$$

Using $D = 14,291$ gives the correct allocation of seats.

Company	Number of Patients	Natural Quota $D = 15,275.7143$	Initial Allocation	Modified Quota $D = 14,291$	Final Allocation
East Michigan	85,750	5.6135	5	6.0003	6
Great Lakes	142,450	9.3253	9	9.9678	9
Mid Michigan	68,170	4.4626	4	4.7701	4
Southeast Michigan	238,280	15.5986	15	16.6734	16
Total	534,650		33		35

(d)

Company	Number of Patients	Natural Quota	Initial & Final Allocation
East Michigan	85,750	5.6135	6
Great Lakes	142,450	9.3253	9
Mid Michigan	68,170	4.4626	4
Southeast Michigan	238,280	15.5986	16
Total	534,650		35

The initial allocation assigns exactly 35 seats, so it is also the final allocation.

(e)

Company	Number of Patients	Natural Quota	Initial & Final Allocation
East Michigan	85,750	5.6135	6
Great Lakes	142,450	9.3253	9
Mid Michigan	68,170	4.4626	4
Southeast Michigan	238,280	15.5986	16
Total	534,650		35

The cutoff between 4 and 5 seats it is $\sqrt{4 \cdot 5} = \sqrt{20} \approx 4.47214$, and we see that exactly 35 are allocated. Therefore, the initial allocation is also the final allocation.

3. The natural divisor is $4234/225 \approx 18.8178$.

(a)

School	Enrollment	Natural Quota	Initial Allocation	Final Allocation
Canyon Hills	1050	55.7982	55	56
Magnolia	924	49.1024	49	49
Ramona	917	48.7305	48	49
Townsend	841	44.6917	44	45
Woodcrest	502	26.6769	26	26
Total	4234		222	225

(b)

School	Enrollment	Natural Quota	Initial Allocation	Relative Fractional Part	Final Allocation
Canyon Hills	1050	55.7982	55	0.01451	55
Magnolia	924	49.1024	49	0.00209	49
Ramona	917	48.7305	48	0.01522	49
Townsend	841	44.6917	44	0.01572	45
Woodcrest	502	26.6769	26	0.02603	27
Total	4234		222		225

(c)

School	Enrollment	Natural Quota	Initial Allocation
Canyon Hills	1050	55.7982	55
Magnolia	924	49.1024	49
Ramona	917	48.7305	48
Townsend	841	44.6917	44
Woodcrest	502	26.6769	26
Total	4234		222

The initial allocation is three seats short. It appears that Canyon Hills, Ramona, and Townsend will gain seats because they have large populations and large fractional parts. Computing their threshold divisors and that of Woodcrest which appears next most likely to gain a seat we find

$$\text{Canyon Hills' threshold divisor} = \frac{1050}{56} = 18.75,$$

$$\text{Ramona's threshold divisor} = \frac{917}{49} \approx 18.7143,$$

$$\text{Townsend's threshold divisor} = \frac{841}{45} \approx 18.6889,$$

$$\text{Woodcrest's threshold divisor} = \frac{502}{27} \approx 18.5926.$$

Letting $D = 18.68$, we get a correct allocation of seats.

School	Enrollment	Natural Quota $D = 18.8178$	Initial Allocation	Modified Quota $D = 18.68$	Final Allocation
Canyon Hills	1050	55.7982	55	56.2099	56
Magnolia	924	49.1024	49	49.4647	49
Ramona	917	48.7305	48	49.0899	49
Townsend	841	44.6917	44	45.0214	45
Woodcrest	502	26.6769	26	26.8737	26
Total	4234		222		225

(d)

School	Enrollment	Natural Quota	Initial Allocation
Canyon Hills	1050	55.7982	56
Magnolia	924	49.1024	49
Ramona	917	48.7305	49
Townsend	841	44.6917	45
Woodcrest	502	26.6769	27
Total	4234		226

The initial allocation assigns one seat too many. It is not clear which school will lose a seat, except that we can rule out Magnolia. We compute the threshold divisors of the remaining four schools and get

$$\text{Canyon Hills' threshold divisor} = \frac{1050}{55.5} \approx 18.9189,$$

$$\text{Ramona's threshold divisor} = \frac{917}{48.5} \approx 18.9072,$$

$$\text{Townsend's threshold divisor} = \frac{841}{44.5} \approx 18.8989,$$

$$\text{Woodcrest's threshold divisor} = \frac{502}{26.5} \approx 18.9434.$$

Using $D = 18.9$ gives a correct allocation of seats.

School	Enrollment	Natural Quota $D=18.8178$	Initial Allocation	Modified Quota $D=18.9$	Final Allocation
Canyon Hills	1050	55.7982	56	55.5556	56
Magnolia	924	49.1024	49	48.8889	49
Ramona	917	48.7305	49	48.5185	49
Townsend	841	44.6917	45	44.4974	44
Woodcrest	502	26.6769	27	26.5608	27
Total	4234		226		225

(e)

School	Enrollment	Natural Quota	Initial Allocation
Canyon Hills	1050	55.7982	56
Magnolia	924	49.1024	49
Ramona	917	48.7305	49
Townsend	841	44.6917	45
Woodcrest	502	26.6769	27
Total	4234		226

The initial allocation assigns one seat too many. As with Webster's method, it is not easy to see which school will lose a seat except that Magnolia can be ruled out. So, we compute the threshold divisors for the remaining four schools and have

$$\text{Canyon Hills' threshold divisor} = \frac{1050}{\sqrt{55 \cdot 56}} = \frac{1050}{\sqrt{3080}} \approx 18.9197,$$

$$\text{Ramona's threshold divisor} = \frac{917}{\sqrt{48 \cdot 49}} = \frac{917}{\sqrt{2352}} \approx 18.9082,$$

$$\text{Townsend's threshold divisor} = \frac{841}{\sqrt{44 \cdot 45}} = \frac{841}{\sqrt{1980}} \approx 18.9001,$$

$$\text{Woodcrest's threshold divisor} = \frac{502}{\sqrt{26 \cdot 27}} = \frac{502}{\sqrt{702}} \approx 18.9468.$$

Letting $D = 18.901$ and using the fact that the cutoff between 44 and 45 seats is $\sqrt{44 \cdot 45} = \sqrt{1980} \approx 44.49719$, we get a correct allocation of seats.

School	Enrollment	Natural Quota $D=18.8178$	Initial Allocation	Modified Quota $D=18.901$	Final Allocation
Canyon Hills	1050	55.7982	56	55.5526	56
Magnolia	924	49.1024	49	48.8863	49
Ramona	917	48.7305	49	48.5160	49
Townsend	841	44.6917	45	44.4950	44
Woodcrest	502	26.6769	27	26.5594	27
Total	4234		226		225

(f) Lowndes' method is the least favorable to the largest school, Canyon Hills.

(g) None of the apportionments violate the Quota Property.

5. We want several states just above the 0.5 cutoff so that too many round up with Webster's method. If quotas are small, the cutoffs for the Hill-Huntington method will be enough smaller so that once one of these states rounds down, the others will have fractional parts below 0.5. We begin with the natural quotas and then multiply them by 100 to get integer populations. It is easiest if all have the same integer part. The natural divisor is $1000/10 = 100$ and the Hill-Huntington cutoff is $\sqrt{2 \cdot 3} \approx 2.44949$.

State	Population	Natural Quota	Initial Allocation by Webster's Method	Initial Allocation by Hill-Huntington Method
A	244	2.4400	2	2
B	251	2.5100	3	3
C	252	2.5200	3	3
D	253	2.5300	3	3
Total	1000	10.0000	11	11

Because we have allocated an extra seat by both methods, both require a divisor larger than the natural divisor. Clearly, state B will be the first to lose a seat. It is enough to show that once the divisor is large enough for state B to lose a seat using the Hill-Huntington method, then state C will also lose a seat by Webster's method, so that any divisor that works for the Hill-Huntington method is too large for Webster's method. We do this by computing threshold divisors.

$$\text{Hill-Huntington threshold divisor for 3 seats for B} = \frac{251}{\sqrt{2 \cdot 3}} = \frac{251}{\sqrt{6}} \approx 102.4703,$$

$$\text{Webster threshold divisor for 3 seats for C} = \frac{252}{2.5} \approx 100.8.$$

(The final apportionment will be 2 seats to A and B and 3 seats to C and D by both methods.)

7. Any final divisor for Jefferson's method is at most the natural divisor (and must be smaller unless all natural quotas are integers). Therefore, a state's final allocation is at least its initial allocation, which in turn is its natural quota rounded down.

CHAPTER 3
Student Solution Manual

SECTION 3.1

1.
$$x^9 = 568$$
$$(x^9)^{1/9} = 568^{1/9}$$
$$x = 568^{1/9} \approx 2.02319948$$

3.
$$y^{50} = 7.23$$
$$(y^{50})^{1/50} = 7.23^{1/50}$$
$$y = 7.23^{1/50} \approx 1.04035789$$

5.
$$x^{12} = 1570$$
$$(x^{12})^{1/12} = 1570^{1/12}$$
$$x = 1570^{1/12} \approx 1.84639651$$

7.
$$3^x = 782$$
$$\log 3^x = \log 782$$
$$x \log 3 = \log 782$$
$$x = \frac{\log 782}{\log 3} \approx 6.06388151$$

9.
$$(1.73)^{-z} = 8$$
$$\log(1.73)^{-z} = \log 8$$
$$-z \log(1.73) = \log 8$$
$$z = \frac{\log 8}{-\log(1.73)} \approx -3.79376085$$

11.
$$9^t = 0.5$$
$$\log 9^t = \log(0.5)$$
$$t \log 9 = \log(0.5)$$
$$t = \frac{\log(0.5)}{\log 9} \approx -0.31546488$$

13.
$$8^{2x} = 49$$
$$\log 8^{2x} = \log 49$$
$$2x \log 8 = \log 49$$
$$x = \frac{\log 49}{2 \log 8} \approx 0.93578497$$

15.
$$(7.53)^{-4x} = 0.249$$
$$\log(7.53)^{-4x} = \log(0.249)$$
$$-4x \log(7.53) = \log(0.249)$$
$$x = \frac{\log(0.249)}{-4 \log(7.53)} \approx 0.17216130$$

17.
$$(1 + x)^{108} = 7.3$$
$$[(1 + x)^{108}]^{1/108} = (7.3)^{1/108}$$
$$1 + x = (7.3)^{1/108}$$
$$x = (7.3)^{1/108} - 1 \approx 0.01857668$$

19.
$$(3 + t)^4 = 82$$
$$[(3 + t)^4]^{1/4} = 82^{1/4}$$
$$3 + t = 82^{1/4}$$
$$t = 82^{1/4} - 3 \approx 0.00921670$$

21.
$$4 + 2^t = 6520$$
$$2^t = 6520 - 4$$
$$2^t = 6516$$
$$\log 2^t = \log 6516$$
$$t \log 2 = \log 6516$$
$$t = \frac{\log 6516}{\log 2} \approx 12.66977089$$

23.
$$4^{-12z} + 1 = 1.75$$
$$4^{-12z} = 1.75 - 1$$
$$4^{-12z} = 0.75$$
$$\log 4^{-12z} = \log(0.75)$$
$$-12z \log 4 = \log(0.75)$$
$$z = \frac{\log(0.75)}{-12 \log 4} \approx 0.01729323$$

25.
$$(1.5)^{2x} - 3 = 8.6$$
$$(1.5)^{2x} = 8.6 + 3$$
$$(1.5)^{2x} = 11.6$$
$$\log(1.5)^{2x} = \log(11.6)$$
$$2x \log(1.5) = \log(11.6)$$
$$x = \frac{\log(11.6)}{2 \log(1.5)} \approx 3.02246118$$

27.
$$186 = \frac{(1.02)^{4x} - 1}{0.02}$$
$$(0.02)(186) = (1.02)^{4x} - 1$$
$$3.72 = (1.02)^{4x} - 1$$
$$3.72 + 1 = (1.02)^{4x}$$
$$4.72 = (1.02)^{4x}$$
$$\log(4.72) = \log(1.02)^{4x}$$
$$\log(4.72) = 4x \log(1.02)$$
$$x = \frac{\log(4.72)}{4 \log(1.02)} \approx 19.59094589$$

29.

$$72.3 = \frac{1 - (1.01)^{-12y}}{0.01}$$

$$(72.3)(0.01) = 1 - (1.01)^{-12y}$$

$$0.723 = 1 - (1.01)^{-12y}$$

$$0.723 - 1 = -(1.01)^{-12y}$$

$$-0.277 = -(1.01)^{-12y}$$

$$(-1)(-0.277) = (-1)(-(1.01)^{-12y})$$

$$0.277 = (1.01)^{-12y}$$

$$\log(0.277) = \log(1.01)^{-12y}$$

$$\log(0.277) = -12y\log(1.01)$$

$$y = \frac{\log(0.277)}{-12\log(1.01)} \approx 10.75121514$$

SECTION 3.2

1. $I = (4000)(0.05)(3) = \$600$

3. $I = (752)(0.121)(8) \approx \$727.94$

5. $I = (35,200)(0.09)(\frac{7}{12}) \approx (35,200)(0.09)(0.58333333) \approx \$1848$

7. $F = 320(1 + (0.03)4) = 320(1.12) = \$358.40$

9. $F = 51,225(1 + (0.0435)(\frac{50}{365})) = 51,225(1 + (0.0435)(0.13698630)) \approx 51,225(1.00595890) \approx \$51,530.24$

11.

$$2000 = P(1 + (0.09)(\tfrac{6}{12}))$$

$$2000 = P(1 + (0.09)(0.5))$$

$$2000 = P(1.045)$$

$$P = \frac{2000}{1.045} \approx \$1913.88$$

13.

$$90,000 = P(1 + (0.142)4)$$

$$90,000 = P(1.568)$$

$$P = \frac{90,000}{1.568} \approx \$57,397.96$$

15.

$$400 = 300(1 + (0.062)t)$$

$$\frac{400}{300} = 1 + (0.062)t$$

$$1.33333333 \approx 1 + (0.062)t$$

$$1.33333333 - 1 \approx (0.062)t$$

$$0.33333333 \approx (0.062)t$$

$$t \approx \frac{0.33333333}{0.062} \approx 5.38 \text{ years}$$

17.

$$140 = 123(1 + r(\tfrac{10}{12}))$$
$$140 \approx 123(1 + r(0.83333333))$$
$$\frac{140}{123} \approx 1 + r(0.83333333)$$
$$1.13821138 \approx 1 + r(0.83333333)$$
$$1.13821138 - 1 \approx r(0.83333333)$$
$$0.13821138 \approx r(0.83333333)$$
$$r \approx \frac{0.13821138}{0.83333333} \approx 0.1659 = 16.59\%$$

19. Substituting $F = 3245$, $t = 6$, and $r = 0.075$ into the simple interest future value formula and solving for P we have

$$3245 = P(1 + (0.075)6)$$
$$3245 = P(1.45)$$
$$P = \frac{3245}{1.45} \approx \$2237.93.$$

So she originally invested $2237.93.

21. In this case $t = 1997 - 1988 = 9$, $P = 6000$, and $F = 49{,}375$. Substituting into the simple interest future value formula and solving for r gives

$$49{,}375 = 6000(1 + r(9))$$
$$\frac{49{,}375}{6000} = 1 + r(9)$$
$$8.22916667 \approx 1 + r(9)$$
$$8.22916667 - 1 \approx r(9)$$
$$7.22916667 \approx r(9)$$
$$r \approx \frac{7.22916667}{9} \approx 0.8032 = 80.32\%.$$

So, figured as an annual simple interest rate, the rate of return was 80.32%.

23. Letting $P = 1000$, $F = 1150$, and $r = 0.0825$ in the simple interest future value formula and solving for t gives

$$1150 = 1000(1 + (0.0825)t)$$
$$\frac{1150}{1000} = 1 + (0.0825)t$$
$$1.15 = 1 + (0.0825)t$$
$$1.15 - 1 = (0.0825)t$$
$$0.15 = (0.0825)t$$
$$t = \frac{0.15}{0.0825} \approx 1.82 \text{ years.}$$

So he must wait 1.82 years until he has enough money for the down payment.

25. Substituting $P = 3000$, $r = 0.48$, and $t = 3$ into the simple interest formula and evaluating I we have

$$I = (3000)(0.48)(3) = 4320 \text{ talents.}$$

Therefore, Brutus would make 4320 talents in interest.

27. Letting $P = 50$, $r = 3$, and $t = \frac{7}{12} \approx 0.58333333$ in the simple interest future value formula and evaluating F, we have

$$F \approx 50(1 + (3)(0.58333333)) = 50(2.74999999) \approx \$137.50.$$

So you would owe \$137.50 if you paid back the loan after 7 months.

29. Letting $P = 500$, $r = 0.06$, and $t = 2$ in the simple interest future value formula and evaluating F we have

$$F = 500(1 + (0.06)2) = 500(1.12) = \$560.$$

Therefore, the man will owe his friend \$560 at the end of the 2 years.

31. Using the simple interest future value formula with $P = 4$, $F = 5$, and $t = \frac{1}{52} \approx 0.01923077$ and solving for r we have

$$5 \approx 4(1 + r(0.01923077))$$

$$\frac{5}{4} \approx 1 + r(0.01923077)$$

$$1.25 \approx 1 + r(0.01923077)$$

$$1.25 - 1 \approx r(0.01923077)$$

$$0.25 \approx r(0.01923077)$$

$$r \approx \frac{0.25}{0.01923077} \approx 13.0000 = 1300\%.$$

We see that the loan sharks were charging 1300% interest.

SECTION 3.3

1. $F = 6000(1 + 0.04)^7 = 6000(1.04)^7 \approx \$7895.59$

3. $F = 7250(1 + \frac{0.10}{4})^{4 \cdot 3} = 7250(1.025)^{12} \approx \$9750.44$

5. $F = 2356(1 + \frac{0.057}{12})^{12 \cdot (9/12)} = 2356(1.00475)^9 \approx \$2458.65$

7. $F = 544(1 + \frac{0.05}{365})^{365 \cdot 15} \approx 544(1.00013699)^{5475} \approx \$1151.61$

9.
$$5400 = P(1 + 0.065)^4$$
$$5400 = P(1.065)^4$$
$$5400 = P(1.28646635)$$
$$P \approx \frac{5400}{1.28646635} \approx \$4197.54$$

11.
$$185 = P(1 + \frac{0.103}{365})^{365 \cdot 2}$$
$$185 \approx P(1.00028219)^{730}$$
$$185 \approx P(1.22871590)$$
$$P \approx \frac{185}{1.22871590} \approx \$150.56$$

13.
$$2000 = 1200(1 + \tfrac{0.058}{12})^{12t}$$
$$2000 \approx 1200(1.00483333)^{12t}$$
$$\frac{2000}{1200} \approx (1.00483333)^{12t}$$
$$1.66666667 \approx (1.00483333)^{12t}$$
$$\log(1.66666667) \approx \log(1.00483333)^{12t}$$
$$\log(1.66666667) \approx 12t \log(1.00483333)$$
$$t \approx \frac{\log(1.66666667)}{12 \log(1.00483333)} \approx 8.83 \text{ years}$$

15.
$$40{,}000 = 10{,}000(1 + \tfrac{0.09}{4})^{4t}$$
$$40{,}000 = 10{,}000(1.0225)^{4t}$$
$$\frac{40{,}000}{10{,}000} = (1.0225)^{4t}$$
$$4 = (1.0225)^{4t}$$
$$\log 4 = \log(1.0225)^{4t}$$
$$\log 4 = 4t \log(1.0225)$$
$$t = \frac{\log(4)}{4 \log(1.0225)} \approx 15.58 \text{ years}$$

17.
$$4500 = 3000(1 + r)^4$$
$$\frac{4500}{3000} = (1 + r)^4$$
$$1.5 = (1 + r)^4$$
$$1.5^{1/4} = [(1 + r)^4]^{1/4}$$
$$1.5^{1/4} = 1 + r$$
$$1.5^{1/4} - 1 = r$$
$$r \approx 0.1067 = 10.67\%$$

19.
$$227 = 65(1 + \tfrac{r}{365})^{365 \cdot 19}$$
$$227 = 65(1 + \tfrac{r}{365})^{6935}$$
$$\frac{227}{65} = (1 + \tfrac{r}{365})^{6935}$$
$$3.49230769 \approx (1 + \tfrac{r}{365})^{6935}$$
$$(3.49230769)^{1/6935} \approx [(1 + \tfrac{r}{365})^{6935}]^{1/6935}$$
$$(3.49230769)^{1/6935} \approx 1 + \tfrac{r}{365}$$
$$(3.49230769)^{1/6935} - 1 \approx \tfrac{r}{365}$$
$$0.00018034 \approx \tfrac{r}{365}$$
$$r \approx 365(0.00018034) \approx 0.0658 = 6.58\%$$

21. $\text{APY} = (1 + \tfrac{0.047}{12})^{12} - 1 \approx (1.00391667)^{12} - 1 \approx 0.0480 = 4.80\%$

23. $\text{APY} = (1 + \tfrac{0.0395}{365})^{365} - 1 \approx (1.00010822)^{365} - 1 \approx 0.0403 = 4.03\%$

25. **(a)** $\text{APY} = (1 + \tfrac{0.05}{4})^4 - 1 = (1.0125)^4 - 1 \approx 0.0509 = 5.09\%$

(b) $\text{APY} = (1 + \frac{0.05}{12})^{12} - 1 \approx (1.00416667)^{12} - 1 \approx 0.0512 = 5.12\%$

(c) $\text{APY} = (1 + \frac{0.05}{365})^{365} - 1 \approx (1.00013699)^{365} - 1 \approx 0.0513 = 5.13\%$

27. (a) $F = 540(1 + 0.08)^5 = 540(1.08)^5 \approx \$793.44$

(b) $F = 540(1 + \frac{0.08}{4})^{4 \cdot 5} = 540(1.02)^{20} \approx \$802.41$

(c) $F = 540(1 + \frac{0.08}{12})^{12 \cdot 5} \approx 540(1.00666667)^{60} \approx \$804.52$

(d) $F = 540(1 + \frac{0.08}{365})^{365 \cdot 5} \approx 540(1.00021918)^{1825} \approx \$805.55$

29. The APY of the NationsBank certificate of deposit was

$$\text{APY} = (1 + \frac{0.0471}{4})^4 - 1 = (1.011775)^4 - 1 \approx 0.0479 = 4.79\%,$$

and the APY of the Bank of America certificate of deposit was

$$\text{APY} = (1 + \frac{0.0470}{365})^{365} - 1 \approx (1.00012877)^{365} - 1 \approx 0.0481 = 4.81\%,$$

so you should choose the Bank of America certificate of deposit.

31. Substituting $F = 15,000$, $P = 12,000$, $t = 2$, and $n = 4$ into the compound interest formula and solving for r we get

$$15,000 = 12,000(1 + \tfrac{r}{4})^{4 \cdot 2}$$
$$15,000 = 12,000(1 + \tfrac{r}{4})^8$$
$$\frac{15,000}{12,000} = (1 + \tfrac{r}{4})^8$$
$$1.25 = (1 + \tfrac{r}{4})^8$$
$$(1.25)^{1/8} = [(1 + \tfrac{r}{4})^8]^{1/8}$$
$$(1.25)^{1/8} = 1 + \tfrac{r}{4}$$
$$(1.25)^{1/8} - 1 = \tfrac{r}{4}$$
$$0.02828559 \approx \tfrac{r}{4}$$
$$r \approx 4(0.02828559) \approx 0.1131 = 11.31\%.$$

So he would have to earn an interest rate of 11.31% to meet his investment goal.

33. Using the compound interest formula with $F = 1.25$ million, $t = 2003 - 1999 = 4$, $r = 0.075$, and $n = 1$ and solving for P we get

$$1.25 = P(1 + 0.075)^4$$
$$1.25 = P(1.075)^4$$
$$1.25 \approx P(1.33546914)$$
$$P \approx \frac{1.25}{1.33546914} \approx 0.93600066 \text{ million.}$$

The Mariners would need to invest \$936,000.66 or about \$0.936 million in 1999.

35. Letting $P = 10,000$, $t = 5$, $r = 0.06625$, and $n = 12$, we use the compound interest formula to compute the future value of the \$10,0000 5-year certificate at 6.625% compounded monthly and find

$$F = 10,000(1 + \frac{0.06625}{12})^{12 \cdot 5} \approx 10,000(1.00552083)^{60} \approx \$13,914.39.$$

Similarly, the future value of the \$10,000 5-year certificate at 5.642% compounded monthly is

$$F = 10{,}000(1 + \tfrac{0.05642}{12})^{12 \cdot 5} \approx 10{,}000(1.00470167)^{60} \approx \$13{,}250.35.$$

Computing the difference between these future values we get

$$\$13{,}914.39 - \$13{,}250.35 = \$664.04.$$

So we see that a \$10,000 5-year certificate at 6.625% compounded monthly earns \$664.04 more in interest than one at 5.642% compounded monthly.

37. Substituting $F = 18{,}000$, $P = 14{,}000$, $r = 0.07$, and $n = 365$ into the compound interest formula and solving for t we have

$$18{,}000 = 14{,}000(1 + \tfrac{0.07}{365})^{365t}$$
$$18{,}000 \approx 14{,}000(1.00019178)^{365t}$$
$$\frac{18{,}000}{14{,}000} \approx (1.00019178)^{365t}$$
$$1.28571429 \approx (1.00019178)^{365t}$$
$$\log(1.28571429) \approx \log(1.00019178)^{365t}$$
$$\log(1.28571429) \approx 365t \log(1.00019178)$$
$$t \approx \frac{\log(1.28571429)}{365 \log(1.00019178)} \approx 3.59 \text{ years.}$$

We see that the family will have to wait 3.59 years until the account has \$18,000 in it.

39. Letting $P = 16{,}710$, $t = 2010 - 1996 = 14$, $r = 0.075$, and $n = 1$ in the compound interest formula and computing F we get

$$F = 16{,}710(1 + 0.075)^{14} = 16{,}710(1.075)^{14} \approx \$45{,}993.34.$$

So we see that tuition will be \$45,993.34 in the year 2010.

41. Substituting $P = 26{,}000$, $t = 1996 - 1980 = 16$, $r = 0.041$, and $n = 1$ in the compound interest formula and computing F we find

$$F = 26{,}000(1 + 0.041)^{16} = 26{,}000(1.041)^{16} \approx \$49{,}452.13.$$

We see that a \$49,452.13 salary in 1996 would be equivalent to a \$26,000 salary in 1980.

43. Letting $F = 100{,}000$, $t = 4$, $r = 0.045$, and $n = 1$ in the compound interest formula and solving for P we have

$$100{,}000 = P(1 + 0.045)^4$$
$$100{,}000 = P(1.045)^4$$
$$100{,}000 \approx P(1.19251860)$$
$$\frac{100{,}000}{1.19251860} \approx \$83{,}856.13.$$

Therefore, the home would be worth \$83,856.13 today.

45. Let P be the present value of the investment. It doubles in value when the future value is $2P$. Substituting $F = 2P$, $r = 0.0392$, and $n = 365$ into the compound interest formula and solving for t we get

$$2P = P(1 + \tfrac{0.0392}{365})^{365t}$$

$$2P \approx P(1.00010740)^{365t}$$

$$\frac{2P}{P} \approx (1.00010740)^{365t}$$

$$2 \approx (1.00010740)^{365t}$$

$$\log 2 \approx \log(1.00010740)^{365t}$$

$$\log 2 \approx 365t \log(1.00010740)$$

$$t \approx \frac{\log 2}{365 \log(1.00010740)} \approx 17.68 \text{ years.}$$

So it will take 17.68 years for the investment to double in value.

47. To find how much is in the first account after 2 years we use the compound interest formula with $P = 950$, $t = 2$, $r = 0.04$, and $n = 365$ and evaluate F to find

$$F = 950(1 + \tfrac{0.04}{365})^{365 \cdot 2} \approx 950(1.00010959)^{730} \approx \$1029.12.$$

To find the amount in the second account after 6 more years we let $P = 1029.12$, $t = 6$, $r = 0.09$, and $n = 4$ in the compound interest formula and evaluate F to find

$$F = 1029.12(1 + \tfrac{0.09}{4})^{4 \cdot 6} = 1029.12(1.0225)^{24} \approx \$1755.44.$$

Therefore, after the entire 8 years the investment will be worth $1755.44.

49. (a) Substituting $P = 10{,}000$, $t = 1000$, and $r = 0.0633$ into the simple interest future value formula and evaluating F we get

$$F = 10{,}000(1 + (0.0633)(1000)) = 10{,}000(64.3) = \$643{,}000.$$

So, we see that the bond would be worth $643,000 in 2861 if simple interest of 6.33% is paid.

(b) Letting $P = 10{,}000$, $t = 1000$, $r = 0.0633$, and $n = 12$ in the compound interest formula and evaluating F we get

$$F = 10{,}000(1 + \tfrac{0.0633}{12})^{12 \cdot 1000} = 10{,}000(1.005275)^{12{,}000} \approx \$2.62172775 \cdot 10^{31}.$$

Therefore, if the interest is compounded monthly with an annual rate of 6.33% then the bond would be worth $2.62172775 \cdot 10^{31}$ in 2861.

51. First, we compute the future value of the first deposit by letting $P = 9200$, $t = 6$, $r = 0.064$, and $n = 4$ in the compound interest formula and evaluating F to find

$$F = 9200(1 + \tfrac{0.064}{4})^{4 \cdot 6} = 9200(1.016)^{24} \approx \$13{,}465.94.$$

Next, we compute the future value of the second deposit by again using the compound interest formula. This time we let $P = 3300$, $t = 4$, $r = 0.064$, and $n = 4$, and evaluating F we get

$$F = 3300(1 + \tfrac{0.064}{4})^{4 \cdot 4} \approx 3300(1.016)^{16} \approx \$4254.15.$$

We now find the total amount in the account 6 years after the first deposit was made by adding these two future values, and we get

$$\$13{,}465.94 + \$4254.15 = \$17{,}720.09.$$

53. **(a)** Using the compound interest formula with $P = 7528$, $t = 10$, $r = 0.0467$, and $n = 1$ and evaluating F gives

$$F = 7528(1 + 0.0467)^{10} = 7528(1.0467)^{10} \approx \$11{,}882.34.$$

So the tuition and fees would be \$11,882.34 in 2006.

(b) Letting $F = 11{,}882.34$, $P = 5914$, $t = 10$, and $n = 1$ and the compound interest formula and solving for r we find

$$11{,}882.34 = 5914(1 + r)^{10}$$

$$\frac{11{,}882.34}{5914} = (1 + r)^{10}$$

$$2.00918837 \approx (1 + r)^{10}$$

$$(2.00918837)^{1/10} \approx [(1 + r)^{10}]^{1/10}$$

$$(2.00918837)^{1/10} \approx 1 + r$$

$$(2.00918837)^{1/10} - 1 \approx r$$

$$r \approx 0.0723 = 7.23\%.$$

We see that an interest rate of 7.23%, compounded annually, would have been earned.

SECTION 3.4

1. $F = 640 \left(\dfrac{(1 + \frac{0.056}{4})^{4 \cdot 15} - 1}{\frac{0.056}{4}} \right) = 640 \left(\dfrac{(1.014)^{60} - 1}{0.014} \right) \approx 640(93.06512658) \approx \$59{,}561.68$

3. $F = 700 \left(\dfrac{(1 + \frac{0.11}{12})^{12 \cdot 10} - 1}{\frac{0.11}{12}} \right) \approx 700 \left(\dfrac{(1.00916667)^{120} - 1}{0.00916667} \right) \approx 700(216.99818885) \approx \$151{,}898.73$

5.

$$40{,}000 = D \left(\frac{(1 + \frac{0.05}{4})^{4 \cdot 11} - 1}{\frac{0.05}{4}} \right)$$

$$40{,}000 = D \left(\frac{(1.0125)^{44} - 1}{0.0125} \right)$$

$$40{,}000 \approx D(58.18833687)$$

$$D \approx \frac{40{,}000}{58.18833687} \approx \$687.42$$

7.

$$5000 = D \left(\frac{(1 + \frac{0.08}{12})^{12 \cdot (21/12)} - 1}{\frac{0.08}{12}} \right)$$

$$5000 \approx D \left(\frac{(1.00666667)^{21} - 1}{0.00666667} \right)$$

$$5000 \approx D(22.46092613)$$

$$D \approx \frac{5000}{22.46092613} \approx \$222.61$$

9.

$$1200 = 75\left(\frac{(1+\frac{0.06}{12})^{12t}-1}{\frac{0.06}{12}}\right)$$

$$1200 = 75\left(\frac{(1.005)^{12t}-1}{0.005}\right)$$

$$\frac{1200}{75} = \left(\frac{(1.005)^{12t}-1}{0.005}\right)$$

$$16 = \left(\frac{(1.005)^{12t}-1}{0.005}\right)$$

$$(16)(0.005) = (1.005)^{12t}-1$$

$$0.08 = (1.005)^{12t}-1$$

$$0.08 + 1 = (1.005)^{12t}$$

$$1.08 = (1.005)^{12t}$$

$$\log(1.08) = \log(1.005)^{12t}$$

$$\log(1.08) = 12t\log(1.005)$$

$$t = \frac{\log(1.08)}{12\log(1.005)} \approx 1.29 \text{ years}$$

11.

$$4000 = 130\left(\frac{(1+\frac{0.12}{4})^{4t}-1}{\frac{0.12}{4}}\right)$$

$$4000 = 130\left(\frac{(1.03)^{4t}-1}{0.03}\right)$$

$$\frac{4000}{130} = \left(\frac{(1.03)^{4t}-1}{0.03}\right)$$

$$30.76923077 \approx \left(\frac{(1.03)^{4t}-1}{0.03}\right)$$

$$(30.76923077)(0.03) \approx (1.03)^{4t}-1$$

$$0.92307692 \approx (1.03)^{4t}-1$$

$$0.92307692 + 1 \approx (1.03)^{4t}$$

$$1.92307692 \approx (1.03)^{4t}$$

$$\log(1.92307692) \approx \log(1.03)^{4t}$$

$$\log(1.92307692) \approx 4t\log(1.03)$$

$$t \approx \frac{\log(1.92307692)}{4\log(1.03)} \approx 5.53 \text{ years}$$

13. (a) Letting $D = 75$, $t = 10$, $r = 0.09$, and $n = 12$ in the systematic savings formula and evaluating F we get

$$F = 75\left(\frac{(1+\frac{0.09}{12})^{12\cdot10}-1}{\frac{0.09}{12}}\right) = 75\left(\frac{(1.0075)^{120}-1}{0.0075}\right) \approx 75(193.51427708) \approx \$14{,}513.57.$$

The account will be worth \$14,513.57 after 10 years.

(b) The amount of the future value that will be from deposits is given by

$$(\$75)(12)(10) = \$9000.$$

(c) The amount of the future value that will be from interest is given by

$$\$14{,}513.57 - \$9000 = \$5513.57.$$

15. Substituting $F = 7000$, $D = 600$, $r = 0.09$, and $n = 4$ into the systematic savings formula and solving for t we get

$$7000 = 600\left(\frac{(1 + \frac{0.09}{4})^{4t} - 1}{\frac{0.09}{4}}\right)$$

$$7000 = 600\left(\frac{(1.0225)^{4t} - 1}{0.0225}\right)$$

$$\frac{7000}{600} = \left(\frac{(1.0225)^{4t} - 1}{0.0225}\right)$$

$$11.66666667 \approx \left(\frac{(1.0225)^{4t} - 1}{0.0225}\right)$$

$$(11.66666667)(0.0225) \approx (1.0225)^{4t} - 1$$

$$0.26250000 \approx (1.0225)^{4t} - 1$$

$$0.26250000 + 1 \approx (1.0225)^{4t}$$

$$1.26250000 \approx (1.0225)^{4t}$$

$$\log(1.26250000) \approx \log(1.0225)^{4t}$$

$$\log(1.26250000) \approx 4t\log(1.0225)$$

$$t \approx \frac{\log(1.26250000)}{4\log(1.0225)} \approx 2.62 \text{ years.}$$

We see that it will take the family 2.62 years to save \$7000.

17. **(a)** Using the systematic savings formula with $D = 100$, $t = 70 - 22 = 48$, $r = 0.07$, and $n = 12$ to compute F we find

$$F = 100\left(\frac{(1 + \frac{0.07}{12})^{12\cdot48} - 1}{\frac{0.07}{12}}\right) \approx 100\left(\frac{(1.00583333)^{576} - 1}{0.00583333}\right)$$

$$\approx 100(4715.91048632) \approx \$471{,}591.05.$$

So he will have \$471,591.05 in the account when he retires.

(b) Using the systematic savings formula with $D = 200$, $t = 70 - 40 = 30$, $r = 0.07$, and $n = 12$ to compute F we find

$$F = 200\left(\frac{(1 + \frac{0.07}{12})^{12\cdot30} - 1}{\frac{0.07}{12}}\right) \approx 200\left(\frac{(1.00583333)^{360} - 1}{0.00583333}\right)$$

$$\approx 200(1219.97003291) \approx \$243{,}994.01.$$

So he will have \$243,994.01 in the account when he retires.

19. **(a)** Using the systematic savings formula with $F = 150{,}000$, $t = 18$, $r = 0.09$, and $n = 12$ and solving for D we have

$$150,000 = D\left(\frac{(1 + \frac{0.09}{12})^{12 \cdot 18} - 1}{\frac{0.09}{12}}\right)$$

$$150,000 = D\left(\frac{(1.0075)^{216} - 1}{0.0075}\right)$$

$$150,000 \approx D(536.35167405)$$

$$D \approx \frac{150,000}{536.35167405} \approx \$279.67.$$

Therefore, deposits of $279.67 should be made.

(b) Letting $F = 150,000$, $t = 8$, $r = 0.09$, and $n = 12$ in the systematic savings formula and solving for D we get

$$150,000 = D\left(\frac{(1 + \frac{0.09}{12})^{12 \cdot 8} - 1}{\frac{0.09}{12}}\right)$$

$$150,000 = D\left(\frac{(1.0075)^{96} - 1}{0.0075}\right)$$

$$150,000 \approx D(139.85616377)$$

$$D \approx \frac{150,000}{139.85616377} \approx \$1072.53.$$

So deposits of $1072.53 should be made.

21. To find the future value of the lump sum payment of $9141 after 18 years we use the compound interest formula with $P = 9141$, $t = 18$, $r = 0.08$, and $n = 12$ and evaluate F to get

$$F = 9141(1 + \tfrac{0.08}{12})^{12 \cdot 18} \approx 9141(1.00666667)^{216} \approx \$38,397.48.$$

To calculate the future value of the $185 payments for 5 years after 18 years we first use the systematic savings formula to find the future value after 5 years. Letting $D = 185$, $t = 5$, $r = 0.08$, and $n = 12$, we see that after 5 years the future value is given by

$$F = 185\left(\frac{(1 + \frac{0.08}{12})^{12 \cdot 5} - 1}{\frac{0.08}{12}}\right) \approx 185\left(\frac{(1.00666667)^{60} - 1}{0.00666667}\right)$$

$$\approx 185(73.47686391) \approx \$13,593.22.$$

Because this money continues to earn interest for 13 more years, we can use the compound interest formula with $P = 13,593.22$, $t = 13$, $r = 0.08$, and $n = 12$ to find that the future value after the entire 18 years is given by

$$F = 13,593.22(1 + \tfrac{0.08}{12})^{12 \cdot 13} \approx 13,593.22(1.00666667)^{156} \approx \$38,325.69.$$

To find the future value in the case of the third option of payments of $80 for 18 years, we use the systematic savings formula with $D = 80$, $t = 18$, $r = 0.08$, and $n = 12$ and find that the future value is given by

$$F = 80\left(\frac{(1 + \frac{0.08}{12})^{12 \cdot 18} - 1}{\frac{0.08}{12}}\right) \approx 80\left(\frac{(1.00666667)^{216} - 1}{0.00666667}\right)$$

$$\approx 80(480.08633873) \approx \$38,406.91.$$

We see that the future values, $38,397.48, $38,325.69, and $38,406.91, under the three different options are fairly close in value.

23. We begin by computing the future value of the $130 deposits. The future value after 10 years is given by F in the systematic savings formula with $D = 130$, $t = 10$, $r = 0.054$, and $n = 4$ and we have

$$F = 130 \left(\frac{(1 + \frac{0.054}{4})^{4 \cdot 10} - 1}{\frac{0.054}{4}} \right) = 130 \left(\frac{(1.0135)^{40} - 1}{0.0135} \right)$$

$$\approx 130(52.57918005) \approx \$6835.29.$$

Because this money continues to earn interest for 20 more years, we can find the future value of the $130 deposits after the entire 30 years by using the compound interest formula with $P = 6835.29$, $t = 20$, $r = 0.054$, and $n = 4$ to find that

$$F = 6835.29(1 + \tfrac{0.054}{4})^{4 \cdot 20} = 6835.29(1.0135)^{80} \approx \$19{,}982.84.$$

Next, we compute the future value of the $200 deposits by letting $D = 200$, $t = 20$, $r = 0.054$, and $n = 4$ in the systematic savings formula and calculating F to find

$$F = 200 \left(\frac{(1 + \frac{0.054}{4})^{4 \cdot 20} - 1}{\frac{0.054}{4}} \right) = 200 \left(\frac{(1.0135)^{80} - 1}{0.0135} \right)$$

$$\approx 200(142.48005745) \approx \$28{,}496.01.$$

Adding the future values of the $130 deposits and the $200 deposit, we find that the amount in the account 30 years after it was first opened is given by

$$\$19{,}982.84 + \$28{,}496.01 = \$48{,}478.85.$$

25. (a) Let F be the future value after t years of a systematic savings plan with an annual interest rate r, compounded n times a year, into which deposits of size D are made at the end of every k periods. To find F, we find the future value of the individual deposits and then sum them. The very last deposit was just made so it is worth only D. The next to last deposit has earned k periods worth of interest, so it is worth $D(1 + \frac{r}{n})^k$. The third to last deposit has earned $2k$ periods of interest, so it is worth $D(1 + \frac{r}{n})^{2k}$. Similarly, the fourth to last deposit has earned interest for $3k$ periods, so it is worth $D(1 + \frac{r}{n})^{3k}$. We continue in this way until we see that the first deposit is now worth

$$D(1 + \tfrac{r}{n})^{(\frac{nt}{k} - 1)k}.$$

Summing the future values of all the deposits, we find that

$$F = D + D(1 + \tfrac{r}{n})^k + D(1 + \tfrac{r}{n})^{2k} + D(1 + \tfrac{r}{n})^{3k} + \cdots + D(1 + \tfrac{r}{n})^{(\frac{nt}{k} - 1)k}$$

$$= D \left(1 + (1 + \tfrac{r}{n})^k + [(1 + \tfrac{r}{n})^k]^2 + [(1 + \tfrac{r}{n})^k]^3 + \cdots + [(1 + \tfrac{r}{n})^k]^{(\frac{nt}{k} - 1)} \right).$$

Applying the formula for the sum of a geometric series we have

$$F = D \left(\frac{[(1 + \frac{r}{n})^k]^{\frac{nt}{k}} - 1}{(1 + \frac{r}{n})^k - 1} \right),$$

and simplifying we find that

$$F = D \left(\frac{(1 + \frac{r}{n})^{nt} - 1}{(1 + \frac{r}{n})^k - 1} \right).$$

(b) Using the formula we found in part (a) with $D = 50$, $t = 7$, $r = 0.06$, $n = 12$, and $k = 3$ we have

$$F = 50 \left(\frac{(1 + \frac{0.06}{12})^{12 \cdot 7} - 1}{(1 + \frac{0.06}{12})^3 - 1} \right) = 50 \left(\frac{(1.005)^{84} - 1}{(1.005)^3 - 1} \right)$$

$$\approx 50(34.51842927) \approx \$1725.92.$$

The future value of the account after 7 years is \$1725.92.

27. To find the future value we sum the future values of each of the individual deposits. The last deposit has been in the account for one period, so it is now worth $D(1 + \frac{r}{n})$, where D is the size of the deposits and r is the annual interest rate, compounded n times a year. The next to last deposit is now worth $D(1 + \frac{r}{n})^2$, and the third to last deposit is worth $D(1 + \frac{r}{n})^3$. We continue in this way until we see that the first deposit is now worth $D(1 + \frac{r}{n})^{nt}$. Summing the future values of the individual deposits, we find that the future value, F, of the account is

$$F = D(1 + \tfrac{r}{n}) + D(1 + \tfrac{r}{n})^2 + D(1 + \tfrac{r}{n})^3 + \cdots + D(1 + \tfrac{r}{n})^{nt}$$
$$= D(1 + \tfrac{r}{n}) \left[1 + (1 + \tfrac{r}{n}) + (1 + \tfrac{r}{n})^2 + \cdots + (1 + \tfrac{r}{n})^{nt-1} \right].$$

Applying the formula for the sum of a geometric series, we see that

$$F = D(1 + \tfrac{r}{n}) \left(\frac{(1 + \frac{r}{n})^{nt} - 1}{\frac{r}{n}} \right).$$

An easier solution is given by noting that when deposits are made at the beginning, rather than the end, of a period, each deposit earns one additional period of interest. Therefore, the expression for F in the systematic savings formula should be multiplied by $(1 + \frac{r}{n})$. We obtain

$$F = D(1 + \tfrac{r}{n}) \left(\frac{(1 + \frac{r}{n})^{nt} - 1}{\frac{r}{n}} \right).$$

SECTION 3.5

1.
$$P = 290.62 \left(\frac{1 - (1 + \frac{0.08}{12})^{-12 \cdot 5}}{\frac{0.08}{12}} \right) \approx 290.62 \left(\frac{1 - (1.00666667)^{-60}}{0.00666667} \right)$$

$$\approx 290.62(49.31842868) \approx \$14,332.92.$$

3.
$$P = 90 \left(\frac{1 - (1 + \frac{0.074}{4})^{-4 \cdot 6}}{\frac{0.074}{4}} \right) = 90 \left(\frac{1 - (1.0185)^{-24}}{0.0185} \right)$$

$$\approx 90(19.23928877) \approx \$1731.54.$$

5.
$$105,000 = R \left(\frac{1 - (1 + \frac{0.0775}{12})^{-12 \cdot 20}}{\frac{0.0775}{12}} \right)$$

$$105,000 \approx R \left(\frac{1 - (1.00645833)^{-240}}{0.00645833} \right)$$

$$105,000 \approx (121.81034772)$$

$$R \approx \frac{105,000}{121.81034772} \approx \$862.00$$

7.

$$7430.65 = R \left(\frac{1 - (1 + \frac{0.13}{4})^{-4 \cdot (18/12)}}{\frac{0.13}{4}} \right)$$

$$7430.65 = R \left(\frac{1 - (1.0325)^{-6}}{0.0325} \right)$$

$$7430.65 \approx (5.37258994)$$

$$R \approx \frac{7430.65}{5.37258994} \approx \$1383.07$$

9.

$$16{,}425 = 500 \left(\frac{1 - (1 + \frac{0.09}{12})^{-12t}}{\frac{0.09}{12}} \right)$$

$$16{,}425 = 500 \left(\frac{1 - (1.0075)^{-12t}}{0.0075} \right)$$

$$\frac{16{,}425}{500} = \left(\frac{1 - (1.0075)^{-12t}}{0.0075} \right)$$

$$32.85 = \left(\frac{1 - (1.0075)^{-12t}}{0.0075} \right)$$

$$(32.85)(0.0075) = 1 - (1.0075)^{-12t}$$

$$0.246375 = 1 - (1.0075)^{-12t}$$

$$0.246375 - 1 = -(1.0075)^{-12t}$$

$$-0.753625 = -(1.0075)^{-12t}$$

$$0.753625 = (1.0075)^{-12t}$$

$$\log(0.753625) = \log(1.0075)^{-12t}$$

$$\log(0.753625) = -12t \log(1.0075)$$

$$t \approx \frac{\log(0.753625)}{-12 \log(1.0075)} \approx 3.15 \text{ years}$$

11.

$$300,000 = 12,000 \left(\frac{1 - (1 + \frac{0.073}{4})^{-4t}}{\frac{0.073}{4}} \right)$$

$$300,000 = 12,000 \left(\frac{1 - (1.01825)^{-4t}}{0.01825} \right)$$

$$\frac{300,000}{12,000} = \left(\frac{1 - (1.01825)^{-4t}}{0.01825} \right)$$

$$25 = \left(\frac{1 - (1.01825)^{-4t}}{0.01825} \right)$$

$$(25)(0.01825) = 1 - (1.01825)^{-4t}$$

$$0.45625 = 1 - (1.01825)^{-4t}$$

$$0.45625 - 1 = -(1.01825)^{-4t}$$

$$-0.54375 = -(1.01825)^{-4t}$$

$$0.54375 = (1.01825)^{-4t}$$

$$\log(0.54375) = \log(1.01825)^{-4t}$$

$$\log(0.54375) = -4t \log(1.01825)$$

$$t = \frac{\log(0.54375)}{-4 \log(1.01825)} \approx 8.42 \text{ years}$$

13. Letting $R = 350$, $t = 8$, $r = 0.09$, and $n = 4$ in the loan formula and evaluating P, we get

$$P = 350 \left(\frac{1 - (1 + \frac{0.09}{4})^{-4 \cdot 8}}{\frac{0.09}{4}} \right) = 350 \left(\frac{1 - (1.0225)^{-32}}{0.0225} \right)$$

$$\approx 350(22.63767419) \approx \$7923.19.$$

So you can afford to borrow $7923.19.

15. (a) Using the loan formula with $P = 9000$, $t = 10$, $r = 0.05$, and $n = 12$ and solving for R we get

$$9000 = R \left(\frac{1 - (1 + \frac{0.05}{12})^{-12 \cdot 10}}{\frac{0.05}{12}} \right)$$

$$9000 \approx R \left(\frac{1 - (1.00416667)^{-120}}{0.00416667} \right)$$

$$9000 \approx R(94.28133295)$$

$$R \approx \frac{9000}{94.28133295} \approx \$95.46.$$

Therefore, the monthly payments are $95.46.

(b) The total amount of payments over the life of the loan is given by

$$(\$95.46)(12)(10) \approx \$11,455.20.$$

(c) Subtracting the principal from the total amount of payments over the life of the loan, we find that the total interest paid over the life of the loan is

$$\$11,455.20 - \$9000 = \$2455.20.$$

17. First, we find the payment amount by letting $P = 9000$, $t = 10$, $r = 0.07$, and $n = 4$ in the loan formula and solving for R to get

$$9000 = R\left(\frac{1 - (1 + \frac{0.07}{4})^{-4 \cdot 10}}{\frac{0.07}{4}}\right)$$

$$9000 = R\left(\frac{1 - (1.0175)^{-40}}{0.0175}\right)$$

$$9000 \approx R(28.59422955)$$

$$R \approx \frac{9000}{28.59422955} \approx \$314.75.$$

To find the balance of the loan after 8 years of payments, we use the loan formula with $R = 314.75$, $t = 10 - 8 = 2$, $r = 0.07$, and $n = 12$ and evaluate P to find that the balance is

$$P = 314.75\left(\frac{1 - (1 + \frac{0.07}{4})^{-4 \cdot 2}}{\frac{0.07}{4}}\right) = 314.75\left(\frac{1 - (1.0175)^{-8}}{0.0175}\right)$$

$$\approx 314.75(7.40505297) \approx \$2330.74.$$

So the student will have to pay $2330.74 to pay off the balance of the loan.

19. The down payment is $(0.10)(\$23,620) = \$2362$, so the principal of the loan is given by

$$\$23,620 - \$2362 = \$21,258.$$

Substituting $P = 21,258$, $R = 2000$, $r = 0.083$, and $n = 4$ into the loan formula and solving for t, we have

$$21,258 = 2000\left(\frac{1 - (1 + \frac{0.083}{4})^{-4t}}{\frac{0.083}{4}}\right)$$

$$21,258 = 2000\left(\frac{1 - (1.02075)^{-4t}}{0.02075}\right)$$

$$\frac{21,258}{2000} = \left(\frac{1 - (1.02075)^{-4t}}{0.02075}\right)$$

$$10.629 = \left(\frac{1 - (1.02075)^{-4t}}{0.02075}\right)$$

$$(10.629)(0.02075) = 1 - (1.02075)^{-4t}$$

$$0.22055175 = 1 - (1.02075)^{-4t}$$

$$0.22055175 - 1 = -(1.02075)^{-4t}$$

$$-0.77944825 = -(1.02075)^{-4t}$$

$$0.77944825 = (1.02075)^{-4t}$$

$$\log(0.77944825) = \log(1.02075)^{-4t}$$

$$\log(0.77944825) = -4t \log(1.02075)$$

$$t = \frac{\log(0.77944825)}{-4 \log(1.02075)} \approx 3.03 \text{ years.}$$

It will take the family 3.03 years to pay off the loan.

21. (a) Substituting $P = 66{,}800$, $t = 30$, $r = 0.1663$, and $n = 12$ into the loan formula and solving for R we get

$$66{,}800 = R\left(\frac{1 - \left(1 + \frac{0.1663}{12}\right)^{-12\cdot30}}{\frac{0.1663}{12}}\right)$$

$$66{,}800 \approx R\left(\frac{1 - (1.01385833)^{-360}}{0.01385833}\right)$$

$$66{,}800 \approx R(71.65005628)$$

$$R \approx \frac{66{,}800}{71.65005628} \approx \$932.31.$$

The monthly payments are \$932.31.

(b) Letting $R = 932.31$, $t = 30 - 5 = 25$, $r = 0.1663$, and $n = 12$ in the loan formula and evaluating P, we get

$$P = 932.31\left(\frac{1 - \left(1 + \frac{0.1663}{12}\right)^{-12\cdot25}}{\frac{0.1663}{12}}\right) \approx 932.31\left(\frac{1 - (1.01385833)^{-300}}{0.01385833}\right)$$

$$\approx 932.31(70.99703186) \approx \$66{,}191.24.$$

The balance of the loan in 1986 is \$66,191.24.

(c) Letting $R = 932.31$, $t = 30 - 19 = 11$, $r = 0.1663$, and $n = 12$ in the loan formula and evaluating P, we get

$$P = 932.31\left(\frac{1 - \left(1 + \frac{0.1663}{12}\right)^{-12\cdot11}}{\frac{0.1663}{12}}\right) \approx 932.31\left(\frac{1 - (1.01385833)^{-132}}{0.01385833}\right)$$

$$\approx 932.31(60.42901088) \approx \$56{,}338.57.$$

So the loan balance in the year 2000 is \$56,338.57.

23. The down payment is $(0.10)(\$16{,}738) = \$1673.80$ so the remaining cost of the car is given by

$$\$16{,}738 - \$1673.80 = \$15{,}064.20.$$

For the 0.9% interest option, we can find the payment by letting $P = 15{,}064.20$, $t = 4$, $r = 0.009$, and $n = 12$ in the loan formula and solving for R to get

$$15{,}064.20 = R\left(\frac{1 - \left(1 + \frac{0.009}{12}\right)^{-12\cdot4}}{\frac{0.009}{12}}\right)$$

$$15{,}064.20 = R\left(\frac{1 - (1.00075)^{-48}}{0.00075}\right)$$

$$15{,}064.20 \approx R(47.12892039)$$

$$R \approx \frac{15{,}064.20}{47.12892039} \approx \$319.64.$$

For the \$750 rebate, we can find the payment by letting $P = 15{,}064.20 - 750 = 14{,}314.20$, $t = 4$, $r = 0.0795$, and $n = 12$ in the loan formula and solving for R to find

$$14{,}314.20 = R\left(\frac{1 - (1 + \frac{0.0795}{12})^{-12 \cdot 4}}{\frac{0.0795}{12}}\right)$$

$$14{,}314.20 = R\left(\frac{1 - (1.006625)^{-48}}{0.006625}\right)$$

$$14{,}314.20 \approx R(41.00131941)$$

$$R \approx \frac{14{,}314.20}{41.00131941} \approx \$349.12.$$

We see that the monthly payments are lowest with the 0.9% interest option.

25. **(a)** Using the loan formula with $P = 138{,}542.91$, $t = 30$, $r = 0.084$, and $n = 12$ and solving for R we get

$$138{,}542.91 = R\left(\frac{1 - (1 + \frac{0.084}{12})^{-12 \cdot 30}}{\frac{0.084}{12}}\right)$$

$$138{,}542.91 = R\left(\frac{1 - (1.007)^{-360}}{0.007}\right)$$

$$138{,}542.91 \approx R(131.26156061)$$

$$R \approx \frac{138{,}542.91}{131.26156061} \approx \$1055.47.$$

The scheduled payment is $1055.47.

(b) Letting $P = 138{,}542.91$, $R = 1055.47 + 200 = 1255.47$, $r = 0.084$, and $n = 12$ in the loan formula and solving for t, we get

$$138{,}542.91 = 1255.47\left(\frac{1 - (1 + \frac{0.084}{12})^{-12t}}{\frac{0.084}{12}}\right)$$

$$138{,}542.91 = 1255.47\left(\frac{1 - (1.007)^{-12t}}{0.007}\right)$$

$$\frac{138{,}542.91}{1255.47} = \left(\frac{1 - (1.007)^{-12t}}{0.007}\right)$$

$$110.35143014 \approx \left(\frac{1 - (1.007)^{-12t}}{0.007}\right)$$

$$(110.35143014)(0.007) \approx 1 - (1.007)^{-12t}$$

$$0.77246001 \approx 1 - (1.007)^{-12t}$$

$$0.77246001 - 1 \approx -(1.007)^{-12t}$$

$$-0.22753999 \approx -(1.007)^{-12t}$$

$$0.22753999 \approx (1.007)^{-12t}$$

$$\log(0.22753999) \approx \log(1.007)^{-12t}$$

$$\log(0.22753999) \approx -12t \log(1.007)$$

$$t \approx \frac{\log(0.22753999)}{-12 \log(1.007)} \approx 17.69 \text{ years.}$$

So it will take 17.69 years to pay off the loan.

(c) By paying $200 extra each month they will save

$$(\$1055.47)(12)(30) - (\$1255.47)(12)(17.69)$$
$$\approx \$379,969.20 - \$266,511.17 = \$113,458.03.$$

27. (a) Substituting $P = 100,000$, $t = 30$, $r = 0.079$, and $n = 12$ into the loan formula and solving for R, we find

$$100,000 = R\left(\frac{1 - (1 + \frac{0.079}{12})^{-12\cdot30}}{\frac{0.079}{12}}\right)$$

$$100,000 \approx R\left(\frac{1 - (1.00658333)^{-360}}{0.00658333}\right)$$

$$100,000 \approx R(137.58846248)$$

$$R \approx \frac{100,000}{137.58846248} \approx \$726.81.$$

So the monthly payment is $726.81.

(b) Substituting $P = 100,000$, $t = 15$, $r = 0.0754$, and $n = 12$ into the loan formula and solving for R, we get

$$100,000 = R\left(\frac{1 - (1 + \frac{0.0754}{12})^{-12\cdot15}}{\frac{0.0754}{12}}\right)$$

$$100,000 \approx R\left(\frac{1 - (1.00628333)^{-180}}{0.00628333}\right)$$

$$100,000 \approx R(107.60942331)$$

$$R \approx \frac{100,000}{107.60942331} \approx \$929.29.$$

Therefore, the monthly payment $929.29.

(c) By choosing the 15-year mortgage rather than the 30-year mortgage, the borrower would save

$$(\$726.81)(12)(30) - (\$929.29)(12)(15)$$
$$\approx \$261,651.60 - \$167,272.20 = \$94,379.40$$

in total payments over the life of the loan.

29. We first find the payment amount by letting $P = 8000$, $t = 10$, $r = 0.06$, and $n = 4$ in the loan formula and solving for R to get

$$8000 = R\left(\frac{1 - (1 + \frac{0.06}{4})^{-4\cdot10}}{\frac{0.06}{4}}\right)$$

$$8000 = R\left(\frac{1 - (1.015)^{-40}}{0.015}\right)$$

$$8000 \approx R(29.91584520)$$

$$R \approx \frac{8000}{29.91584520} \approx \$267.42.$$

Payment Number	Payment	Interest Paid	Principal Paid	Balance
				8000.00
1	267.42	120.00	147.42	7852.58
2	267.42	117.79	149.63	7702.95
3	267.42	115.54	151.88	7551.07
4	267.42	113.27	154.15	7396.92

31. First, we find the payment amount by substituting $P = 5429.39$, $t = \frac{6}{12}$, $r = 0.084$, and $n = 12$ into the loan formula and solving for R to find

$$5429.39 = R\left(\frac{1 - (1 + \frac{0.084}{12})^{-12\cdot(6/12)}}{\frac{0.084}{12}}\right)$$

$$5429.39 = R\left(\frac{1 - (1.007)^{-6}}{0.007}\right)$$

$$5429.39 \approx R(5.85570138)$$

$$R \approx \frac{5429.39}{5.85570138} \approx \$927.20.$$

Payment Number	Payment	Interest Paid	Principal Paid	Balance
				5429.38
1	927.20	38.01	889.19	4540.19
2	927.20	31.78	895.42	3644.77
3	927.20	25.51	901.69	2743.08
4	927.20	19.20	908.00	1835.08
5	927.20	12.85	914.35	920.73
6	927.18	6.45	920.73	0

33. We find the payment amount by letting $P = 7300$, $t = \frac{9}{12}$, $r = 0.115$, and $n = 4$ in the loan formula and solving for R to get

$$7300 = R\left(\frac{1 - (1 + \frac{0.115}{4})^{-4\cdot(9/12)}}{\frac{0.115}{4}}\right)$$

$$7300 = R\left(\frac{1 - (1.02875)^{-3}}{0.02875}\right)$$

$$7300 \approx R(2.83542299)$$

$$R \approx \frac{7300}{2.83542299} \approx \$2574.57.$$

Payment Number	Payment	Interest Paid	Principal Paid	Balance
				7300.00
1	2574.57	209.88	2364.69	4935.31
2	2574.57	141.89	2432.68	2502.63
3	2574.58	71.95	2502.63	0

35. **(a)** First, we find the payment amount by letting $P = 115,000$, $t = 30$, $r = 0.0945$, and $n = 12$ in the loan formula and solving for R to get

$$115,000 = R \left(\frac{1 - (1 + \frac{0.0945}{12})^{-12 \cdot 30}}{\frac{0.0945}{12}} \right)$$

$$115,000 = R \left(\frac{1 - (1.007875)^{-360}}{0.007875} \right)$$

$$115,000 \approx R(119.44454571)$$

$$R \approx \frac{115,000}{119.44454571} \approx \$962.79.$$

Therefore, the total amount the man would make in payments over the life of the loan for the original loan would be

$$(\$962.79)(12)(30) = \$346,604.40.$$

(b) The principal of the new loan would be the balance of the original loan after 10 years of payments. To find this balance, we let $R = 962.79$, $t = 30 - 10 = 20$, $r = 0.0945$, and $n = 12$ in the loan formula and evaluate P to get

$$P = 962.79 \left(\frac{1 - (1 + \frac{0.0945}{12})^{-12 \cdot 20}}{\frac{0.0945}{12}} \right) = 962.79 \left(\frac{1 - (1.007875)^{-240}}{0.007875} \right)$$

$$\approx 962.79(107.65786049) \approx \$103,651.91.$$

So the principal of the new loan would be $103,651.91.

(c) We first find the payment amount for the second loan by letting $P = 103,651.91$, $t = 20$, $r = 0.085$, and $n = 12$ in the loan formula and solving for R to get

$$103,651.91 = R \left(\frac{1 - (1 + \frac{0.085}{12})^{-12 \cdot 20}}{\frac{0.085}{12}} \right)$$

$$103,651.91 \approx R \left(\frac{1 - (1.00708333)^{-240}}{0.00708333} \right)$$

$$103,651.91 \approx R(115.23087344)$$

$$R \approx \frac{103,651.91}{115.23087344} \approx \$899.52.$$

The total amount the man will make in payments for both loans if he decides to refinance is given by

$$(\$962.79)(12)(10) + (\$899.52)(12)(20) = \$331,419.60.$$

(d) The total amount the man will save in payments by refinancing is given by

$$\$346,604.40 - \$331,419.60 = \$15,184.80,$$

so the man should refinance.

37. **(a)** We find the payment amount in the case of the 7.625% loan by letting $P = 95,000$, $t = 30$, $r = 0.07625$, and $n = 12$ in the loan formula and solving for R to get

$$95{,}000 = R\left(\frac{1-\left(1+\frac{0.07625}{12}\right)^{-12\cdot 30}}{\frac{0.07625}{12}}\right)$$

$$95{,}000 \approx R\left(\frac{1-(1.00635417)^{-360}}{0.00635417}\right)$$

$$95{,}000 \approx R(141.28404163)$$

$$R \approx \frac{95{,}000}{141.28404163} \approx \$672.40.$$

To find the payment amount in the case of the 7.25% loan, we let $P = 95{,}000$, $t = 30$, $r = 0.0725$, and $n = 12$ in the loan formula and solve for R, getting

$$95{,}000 = R\left(\frac{1-\left(1+\frac{0.0725}{12}\right)^{-12\cdot 30}}{\frac{0.0725}{12}}\right)$$

$$95{,}000 \approx R\left(\frac{1-(1.00604167)^{-360}}{0.00604167}\right)$$

$$95{,}000 \approx R(146.58961789)$$

$$R \approx \frac{95{,}000}{146.58961789} \approx \$648.07.$$

The amount saved in payments over the life of the loan by paying the point is given by

$$(\$672.40)(12)(30) - (\$648.07)(12)(30) = \$8758.80.$$

(b) The point costs $(0.01)(\$95{,}000) = \$950$ and the amount saved over the life of the loan is \$8758.80, so the man should pay the point.

(c) The amount saved in payments over 2 years by paying the point is given by

$$(\$672.40)(12)(2) - (\$648.07)(12)(2) = \$583.92.$$

(d) Letting $R = 648.07$, $t = 30 - 2 = 28$, $r = 0.0725$, and $n = 12$ in the loan formula and evaluating P, we get

$$P = 648.07\left(\frac{1-\left(1+\frac{0.0725}{12}\right)^{-12\cdot 28}}{\frac{0.0725}{12}}\right) \approx 648.07\left(\frac{1-(1.00604167)^{-336}}{0.00604167}\right)$$

$$\approx 648.07(143.64571598) \approx \$93{,}092.48.$$

So, if the point is paid, the balance after 2 years is \$93,092.48.

(e) Letting $R = 672.40$, $t = 30 - 2 = 28$, $r = 0.07625$, and $n = 12$ in the loan formula and evaluating P we have

$$P = 672.40\left(\frac{1-\left(1+\frac{0.07625}{12}\right)^{-12\cdot 28}}{\frac{0.07625}{12}}\right) \approx 672.40\left(\frac{1-(1.00635417)^{-336}}{0.00635417}\right)$$

$$\approx 672.40(138.64189320) \approx \$93{,}222.81.$$

So, if the point is not paid, the balance after 2 years is \$93,222.81.

(f) The point costs \$950, and the man saves \$583.92 in payments and pays off $\$93{,}222.81 - \$93{,}092.48 = \$130.33$ more of the loan balance over 2 years. Because

$$\$583.92 + \$130.33 = \$714.25 < \$950,$$

the man should not pay the point.

39. (a) The future value of the point after 30 years can be found by letting $P = 950$, $t = 30$, $r = 0.03$, and $n = 12$ in the compound interest formula and evaluating F to get

$$F = 950(1 + \tfrac{0.03}{12})^{12 \cdot 30} = 950(1.0025)^{360} \approx \$2334.00.$$

We find the future value of the difference in monthly payments by letting $D = 672.40 - 648.07 = 24.33$, $t = 30$, $r = 0.03$, and $n = 12$ in the systematic savings formula and evaluating F to get

$$F = 24.33 \left(\frac{(1 + \tfrac{0.03}{12})^{12 \cdot 30} - 1}{\tfrac{0.03}{12}} \right) = 24.33 \left(\frac{(1.0025)^{360} - 1}{0.0025} \right)$$

$$\approx 24.33(582.73688460) \approx \$14{,}177.99.$$

Because $\$14{,}177.99 > \$2334.00$, we see that the borrower should pay the point if he plans to hold the loan for the entire 30 years.

(b) The future value of the point after 2 years can be found by substituting $P = 950$, $t = 2$, $r = 0.03$, and $n = 12$ into the compound interest formula and evaluating F to find

$$F = 950(1 + \tfrac{0.03}{12})^{12 \cdot 2} = 950(1.0025)^{24} \approx \$1008.67.$$

We find the future value of the difference in monthly payments by substituting $D = 672.40 - 648.07 = 24.33$, $t = 2$, $r = 0.03$, and $n = 12$ into the systematic savings formula and evaluating F to get

$$F = 24.33 \left(\frac{(1 + \tfrac{0.03}{12})^{12 \cdot 2} - 1}{\tfrac{0.03}{12}} \right) = 24.33 \left(\frac{(1.0025)^{24} - 1}{0.0025} \right)$$

$$\approx 24.33(24.70281770) \approx \$601.02.$$

The man pays off $\$130.33$ more of the loan balance over 2 years if he pays the point. Because

$$\$601.02 + \$130.33 = \$731.35 < \$1008.67,$$

the man should not pay the point.

41. Letting $R = 500$, $t = 15$, $r = 0.0725$, and $n = 12$ in the loan formula and evaluating P, we have

$$P = 500 \left(\frac{1 - (1 + \tfrac{0.0725}{12})^{-12 \cdot 15}}{\tfrac{0.0725}{12}} \right) \approx 500 \left(\frac{1 - (1.00604167)^{-180}}{0.00604167} \right)$$

$$\approx 500(109.54545023) \approx \$54{,}772.73.$$

So a lump sum of $\$54{,}772.73$ should be invested at the start.

43. Substituting $P = 257{,}548.72$, $R = 4000$, $r = 0.0672$, and $n = 12$ into the loan formula and solving for t, we get

$$257{,}548.72 = 4000 \left(\frac{1 - (1 + \frac{0.0672}{12})^{-12t}}{\frac{0.0672}{12}} \right)$$

$$257{,}548.72 = 4000 \left(\frac{1 - (1.0056)^{-12t}}{0.0056} \right)$$

$$\frac{257{,}548.72}{4000} = \left(\frac{1 - (1.0056)^{-12t}}{0.0056} \right)$$

$$64.38718 = \left(\frac{1 - (1.0056)^{-12t}}{0.0056} \right)$$

$$(64.38718)(0.0056) = 1 - (1.0056)^{-12t}$$

$$0.36056821 \approx 1 - (1.0056)^{-12t}$$

$$0.36056821 - 1 \approx -(1.0056)^{-12t}$$

$$-0.63943179 \approx -(1.0056)^{-12t}$$

$$0.63943179 \approx (1.0056)^{-12t}$$

$$\log(0.63943179) \approx \log(1.0056)^{-12t}$$

$$\log(0.63943179) \approx -12t \log(1.0056)$$

$$t \approx \frac{\log(0.63943179)}{-12 \log(1.0056)} \approx 6.67 \text{ years.}$$

45. Letting $R = \frac{1{,}000{,}000}{(12 \cdot 30)} \approx 2777.78$, $t = 30$, $r = 0.065$, and $n = 12$ in the loan formula and evaluating P we get

$$P = 2777.78 \left(\frac{1 - (1 + \frac{0.065}{12})^{-12 \cdot 30}}{\frac{0.065}{12}} \right) \approx 2777.78 \left(\frac{1 - (1.00541667)^{-360}}{0.00541667} \right)$$

$$\approx 2777.78(158.21075369) \approx \$439{,}474.67.$$

So a lump sum of \$439,474.67 would need to be invested at the start.

47. **(a)** Letting $P = 3{,}000{,}000$, $t = 24$, $r = 0.0682$, and $n = 1$ in the given formula and solving for R, we get

$$3{,}000{,}000 = R + R \left(\frac{1 - (1.0682)^{-24}}{0.0682} \right)$$

$$3{,}000{,}000 \approx R + R(11.65284418)$$

$$3{,}000{,}000 \approx R(12.65284418)$$

$$R \approx \frac{3{,}000{,}000}{12.65284418} \approx \$237{,}100.84.$$

The amount of each payment is \$237,100.84.

(b) The total amount paid out in payments is

$$(\$237{,}100.84)(25) = \$5{,}927{,}521.$$

49. **(a)** The future value after t years of the principal P earning interest r compounded n times a year is given by

$$F = P(1 + \tfrac{r}{n})^{nt}.$$

Using the result of exercise 25(a) in Section 3.4, we see that the future value of the series of payments of size R made at the end of every k periods for t years earning interest r, compounded n times a year, is given by

$$F = R \left(\frac{(1 + \frac{r}{n})^{nt} - 1}{(1 + \frac{r}{n})^{k} - 1} \right).$$

Setting these future values equal, we have

$$P(1 + \tfrac{r}{n})^{nt} = R \left(\frac{(1 + \frac{r}{n})^{nt} - 1}{(1 + \frac{r}{n})^{k} - 1} \right).$$

Solving for P we get

$$P = \frac{1}{(1 + \frac{r}{n})^{nt}} \left[R \left(\frac{(1 + \frac{r}{n})^{nt} - 1}{(1 + \frac{r}{n})^{k} - 1} \right) \right].$$

Simplifying, we arrive at the following loan formula

$$P = R \left(\frac{1 - (1 + \frac{r}{n})^{-nt}}{(1 + \frac{r}{n})^{k} - 1} \right).$$

(b) Substituting $P = 5000$, $t = 4$, $r = 0.08$, $n = 12$, and $k = 3$ into the loan formula we found in part (a) and solving for R, we get

$$5000 = R \left(\frac{1 - (1 + \frac{0.08}{12})^{-12 \cdot 4}}{(1 + \frac{0.08}{12})^{3} - 1} \right)$$

$$5000 \approx R \left(\frac{1 - (1.00666667)^{-48}}{(1.00666667)^{3} - 1} \right)$$

$$5000 \approx R(13.56334664)$$

$$R \approx \frac{5000}{13.56334664} \approx \$368.64.$$

Therefore, the monthly payment is \$368.64.

Chapter 3 Review Exercises

1. To find the future value of a \$10,000 6-month certificate of deposit earning 15.79% compounded monthly, we let $P = 10,000$, $t = \frac{6}{12}$, $r = 0.1579$, and $n = 12$ in the compound interest formula and evaluate F to get

$$F = 10,000(1 + \tfrac{0.1579}{12})^{12 \cdot (6/12)} \approx 10,000(1.01315833)^{6} \approx \$10,815.93.$$

To find the future value of a \$10,000 6-month certificate of deposit earning 3.28% compounded monthly, we let $P = 10,000$, $t = \frac{6}{12}$, $r = 0.0328$, and $n = 12$ in the compound interest formula and evaluate F to find

$$F = 10,000(1 + \tfrac{0.0328}{12})^{12 \cdot (6/12)} \approx 10,000(1.00273333)^{6} \approx \$10,165.12$$

Therefore, a \$10,000 6-month certificate would earn \$10,815.93 − \$10,165.12 = \$650.81 more in interest at 15.79% compounded monthly than at 3.28% compounded monthly.

3. (a) Using the systematic savings formula with $D = 500$, $t = 68 - 22 = 46$, $r = 0.07$, and $n = 4$ to compute F, we find

$$F = 500 \left(\frac{(1 + \frac{0.07}{4})^{4 \cdot 46} - 1}{\frac{0.07}{4}} \right) = 500 \left(\frac{(1.0175)^{184} - 1}{0.0175} \right)$$

$$\approx 500(1333.75373934) \approx \$666{,}876.87.$$

The account will have \$666,876.87 in it when the man is 68 years old.

(b) The amount of the future value that will be from deposits is given by

$$(\$500)(4)(46) = \$92{,}000.$$

(c) The interest paid over the life of the loan is given by

$$\$666{,}876.87 - \$92{,}000 = \$574{,}876.87.$$

5. Using the compound interest formula with $F = 40{,}000$, $t = 1990 - 1940 = 50$, $r = 0.046$, and $n = 1$ and solving for P, we get

$$40{,}000 = P(1 + 0.046)^{50}$$

$$40{,}000 = P(1.046)^{50}$$

$$40{,}000 \approx P(9.47510926)$$

$$P \approx \frac{40{,}000}{9.47510926} \approx \$4221.59.$$

We see that a \$4221.59 salary in 1940 would be equivalent to a \$40,000 salary in 1990.

7. (a) Letting $R = 600$, $t = 30$, $r = 0.0963$, and $n = 12$ in the loan formula and evaluating P, we get

$$P = 600 \left(\frac{1 - (1 + \frac{0.0963}{12})^{-12 \cdot 30}}{\frac{0.0963}{12}} \right) = 600 \left(\frac{1 - (1.008025)^{-360}}{0.008025} \right)$$

$$\approx 600(117.59788289) \approx \$70{,}558.73.$$

So the family could afford a loan of \$70,558.73.

(b) Letting $R = 600$, $t = 30$, $r = 0.1119$, and $n = 12$ in the loan formula and evaluating P, we find

$$P = 600 \left(\frac{1 - (1 + \frac{0.1119}{12})^{-12 \cdot 30}}{\frac{0.1119}{12}} \right) = 600 \left(\frac{1 - (1.009325)^{-360}}{0.009325} \right)$$

$$\approx 600(103.44388183) \approx \$62{,}066.33.$$

So the family could afford a loan of \$62,066.33.

9. Substituting $P = 30{,}000$, $r = 0.10$, and $t = 1570 - 1561 = 9$ into the simple interest formula and evaluating I, we have

$$I = (30{,}000)(0.10)(9) = 27{,}000 \text{ pounds.}$$

So she would have paid 27,000 pounds in simple interest by 1570.

11. (a) First, we find the payment amount by letting $P = 18{,}493.05$, $t = 5$, $r = 0.099$, and $n = 12$ in the loan formula and solving for R to get

$$18{,}493.05 = R\left(\frac{1 - (1 + \frac{0.099}{12})^{-12 \cdot 5}}{\frac{0.099}{12}}\right)$$

$$18{,}493.05 = R\left(\frac{1 - (1.00825)^{-60}}{0.00825}\right)$$

$$18{,}493.05 \approx R(47.17454194)$$

$$R \approx \frac{18{,}493.05}{47.17454194} \approx \$392.01.$$

Now we find the loan balance after 1 year by letting $R = 392.01$, $t = 5 - 1 = 4$, $r = 0.099$, and $n = 12$ in the loan formula and evaluating P to get

$$P = 392.01\left(\frac{1 - (1 + \frac{0.099}{12})^{-12 \cdot 4}}{\frac{0.099}{12}}\right) = 392.01\left(\frac{1 - (1.00825)^{-48}}{0.00825}\right)$$

$$\approx 392.01(39.50291737) \approx \$15{,}485.54.$$

So, the loan balance after 1 year is \$15,485.54.

(b) Letting $R = 392.01$, $t = 5 - 4 = 1$, $r = 0.099$, and $n = 12$ in the loan formula and evaluating P, we get

$$P = 392.01\left(\frac{1 - (1 + \frac{0.099}{12})^{-12 \cdot 1}}{\frac{0.099}{12}}\right) = 392.01\left(\frac{1 - (1.00825)^{-12}}{0.00825}\right)$$

$$\approx 392.01(11.38052805) \approx \$4461.28.$$

So the loan balance after 4 years is \$4461.28.

13. (a) Letting $P = 14{,}000$, $t = 10$, $r = 0.084$, and $n = 12$ in the loan formula and solving for R, we get

$$14{,}000 = R\left(\frac{1 - (1 + \frac{0.084}{12})^{-12 \cdot 10}}{\frac{0.084}{12}}\right)$$

$$14{,}000 = R\left(\frac{1 - (1.007)^{-120}}{0.007}\right)$$

$$14{,}000 \approx R(81.00346993)$$

$$R \approx \frac{14{,}000}{81.00346993} \approx \$172.83.$$

The scheduled payment is \$172.83.

(b) Using the loan formula with $P = 14{,}000$, $R = 172.83 + 50 = 222.83$, $r = 0.084$, and $n = 12$ and solving for t, we find

$$14{,}000 = 222.83 \left(\frac{1 - (1 + \frac{0.084}{12})^{-12t}}{\frac{0.084}{12}} \right)$$

$$14{,}000 = 222.83 \left(\frac{1 - (1.007)^{-12t}}{0.007} \right)$$

$$\frac{14{,}000}{222.83} = \left(\frac{1 - (1.007)^{-12t}}{0.007} \right)$$

$$62.82816497 \approx \left(\frac{1 - (1.007)^{-12t}}{0.007} \right)$$

$$(62.82816497)(0.007) \approx 1 - (1.007)^{-12t}$$

$$0.43979715 \approx 1 - (1.007)^{-12t}$$

$$0.43979715 - 1 \approx -(1.007)^{-12t}$$

$$-0.56020285 \approx -(1.007)^{-12t}$$

$$0.56020285 \approx (1.007)^{-12t}$$

$$\log(0.56020285) \approx \log(1.007)^{-12t}$$

$$\log(0.56020285) \approx -12t \log(1.007)$$

$$t \approx \frac{\log(0.56020285)}{-12 \log(1.007)} \approx 6.92 \text{ years.}$$

It will take 6.92 years to pay off the loan.

(c)   By paying $50 extra each month the company will save

$$(\$172.83)(12)(10) - (\$222.83)(12)(6.92) \approx \$20{,}739.60 - \$18{,}503.80 \approx \$2235.80.$$

15. $\text{APY} = (1 + \frac{0.0575}{12})^{12} - 1 \approx (1.00479167)^{12} - 1 \approx 0.0590 = 5.90\%$

17.   Substituting $P = 3800$, $R = 200$, $r = 0.05$, and $n = 4$ into the loan formula and solving for t, we find

$$3800 = 200 \left(\frac{1 - (1 + \frac{0.05}{4})^{-4t}}{\frac{0.05}{4}} \right)$$

$$3800 = 200 \left(\frac{1 - (1.0125)^{-4t}}{0.0125} \right)$$

$$\frac{3800}{200} = \left(\frac{1 - (1.0125)^{-4t}}{0.0125} \right)$$

$$19 = \left(\frac{1 - (1.0125)^{-4t}}{0.0125} \right)$$

$$(19)(0.0125) = 1 - (1.0125)^{-4t}$$

$$0.2375 = 1 - (1.0125)^{-4t}$$

$$0.2375 - 1 = -(1.0125)^{-4t}$$

$$-0.7625 = -(1.0125)^{-4t}$$

$$0.7625 = (1.0125)^{-4t}$$

$$\log(0.7625) = \log(1.0125)^{-4t}$$

$$\log(0.7625) = -4t \log(1.0125)$$

$$t = \frac{\log(0.7625)}{-4 \log(1.0125)} \approx 5.46 \text{ years.}$$

Therefore, it will take 5.46 years to pay off the loan.

19. To find the amount in the account after the first 3 years, we let $D = 300$, $t = 3$, $r = 0.045$, and $n = 4$ in the systematic savings formula and evaluate F to find

$$F = 300 \left(\frac{(1 + \frac{0.045}{4})^{4 \cdot 3} - 1}{\frac{0.045}{4}} \right) = 300 \left(\frac{(1.01125)^{12} - 1}{0.01125} \right)$$

$$\approx 300(12.77106140) \approx \$3831.32.$$

To find the amount in the account after the remaining 5 years, we use the compound interest formula with $P = 3831.32$, $t = 5$, $r = 0.04$, and $n = 4$ and evaluate F to find

$$F = 3831.32(1 + \tfrac{0.04}{4})^{4 \cdot 5} = 3831.32(1.01)^{20} \approx \$4674.94.$$

So we see that the account will be worth \$4674.94 8 years after the woman opened it.

CHAPTER 4

Student Solution Manual

SECTION 4.1

1. **(a)** $P(6) = \dfrac{1}{6}$ **(c)** $P(\text{less than } 3) = \dfrac{2}{6} = \dfrac{1}{3}$

   (b) $P(\text{odd}) = \dfrac{3}{6} = \dfrac{1}{2}$ **(d)** $P(7) = \dfrac{0}{6} = 0$

3. $P(\text{female}) = \dfrac{12}{12 + 8} = \dfrac{12}{20} = \dfrac{3}{5}$

5. $P(\text{did not die of cancer}) = 1 - P(\text{die of cancer}) = 1 - 0.234 = 0.766$

7. **(a)** $P(\text{jack}) = \dfrac{4}{52} = \dfrac{1}{13}$ **(b)** $P(\text{face card}) = \dfrac{12}{52} = \dfrac{3}{13}$

9. $P(\text{not an ace}) = 1 - P(\text{ace}) = 1 - \dfrac{4}{52} = 1 - \dfrac{1}{13} = \dfrac{12}{13}$

11. $P(2) = \dfrac{5}{8}$

13. **(a)** H1, H2, H3, H4, H5, H6, T1, T2, T3, T4, T5, T6

    (b) $P(\text{tail and } 3) = \dfrac{1}{12}$

    (c) $P(\text{head and even}) = \dfrac{3}{12} = \dfrac{1}{4}$

15. $P(\text{winning ticket}) = \dfrac{6}{870} = \dfrac{1}{145}$

17. The possible outcomes are BBB, BBG, BGB, BGG, GGG, GGB, GBG, and GBB. Therefore, $P(\text{exactly one girl}) = 3/8$.

19. As we saw in Example 5 of the text, there are 36 possible outcomes. The outcomes that have a sum of 9 are (6, 3), (3, 6), (5, 4), and (4, 5). Therefore, $P(\text{sum of } 9) = 4/36 = 1/9$.

21. The outcomes that have a sum of 6 are (5, 1), (1, 5), (4, 2), (2, 4), and (3, 3). Therefore, $P(\text{sum of } 6) = 5/36$. It follows that $P(\text{not a sum of } 6) = 1 - P(\text{sum of } 6) = 1 - (5/36) = 31/36$.

23. $P(\text{correct answer}) = \dfrac{1}{5}$

25. **(a)** $P = \dfrac{250}{1000} = 0.25$

    (b) $P = \dfrac{250 + 200 + 157}{1000} = \dfrac{607}{1000} = 0.607$

27. The possible outcomes are:

(1, 1)	(1, 2)	(1, 2)	(1, 3)	(1, 3)	(1, 4)
(3, 1)	(3, 2)	(3, 2)	(3, 3)	(3, 3)	(3, 4)
(4, 1)	(4, 2)	(4, 2)	(4, 3)	(4, 3)	(4, 4)
(5, 1)	(5, 2)	(5, 2)	(5, 3)	(5, 3)	(5, 4)
(6, 1)	(6, 2)	(6, 2)	(6, 3)	(6, 3)	(6, 4)
(8, 1)	(8, 2)	(8, 2)	(8, 3)	(8, 3)	(8, 4)

(a) $P(\text{sum of 2}) = \dfrac{1}{36}$

(b) $P(\text{sum of 3}) = \dfrac{2}{36} = \dfrac{1}{18}$

(c) $P(\text{sum of 4}) = \dfrac{3}{36} = \dfrac{1}{12}$

(d) $P(\text{sum of 5}) = \dfrac{4}{36} = \dfrac{1}{9}$

(e) $P(\text{sum of 6}) = \dfrac{5}{36}$

(f) $P(\text{sum of 7}) = \dfrac{6}{36} = \dfrac{1}{6}$

(g) $P(\text{sum of 8}) = \dfrac{5}{36}$

(h) $P(\text{sum of 9}) = \dfrac{4}{36} = \dfrac{1}{9}$

(i) $P(\text{sum of 10}) = \dfrac{3}{36} = \dfrac{1}{12}$

(j) $P(\text{sum of 11}) = \dfrac{2}{36} = \dfrac{1}{18}$

(k) $P(\text{sum of 12}) = \dfrac{1}{36}$

29. $P(\text{field goal}) = \dfrac{21}{26} \approx 0.8077$

31.   (a) $P = \dfrac{757}{1{,}657{,}822{,}000} \approx 0.000000451$

(b) $P = \dfrac{9}{23{,}800{,}000} \approx 0.000000378$

(c)   If every flight had the same number of passengers and everyone died on every flight that involved fatalities, then the probabilities in (a) and (b) would be equal. If there were many survivors of plane accidents involving fatalities, then we should expect the probability in (b) to be larger than the probability in (a). Because the opposite is true, it does not seem likely that there are many survivors of plane accidents involving fatalities.

33.   Between the two cards there are four sides that could be turned up, three of which are blue and one red. Two of the three blue faces also have blue on the reverse side. Therefore, if you are looking at a blue side, the probability that the other side is also blue is 2/3.

SECTION 4.2

1.   There are three ways to roll an even number and three ways not to roll an even number, so the odds against rolling an even number are 3 to 3. Dividing by 3, we express these odds as 1 to 1.

3. $P(\text{AFC wins}) = \dfrac{2}{7+2} = \dfrac{2}{9}$

5.   The odds against the team winning the game are a to b, where

$$\frac{a}{b} = \frac{1 - 0.85}{0.85} = \frac{0.15}{0.85} = \frac{15}{85} = \frac{3}{17}$$

So we see the odds are 3 to 17.

7. $P(\text{winning \$2}) = \dfrac{2}{9+2} = \dfrac{2}{11}$

9. There are four possible outcomes: HH, HT, TH, and TT. We see that there is one way to get two heads and three ways not to get two heads. Therefore, the odds against getting two heads are 3 to 1.

11. The odds against the spinner landing on blue are a to b, where

$$\frac{a}{b} = \frac{1 - (3/8)}{3/8} = \frac{5/8}{3/8} = \frac{5}{3}.$$

So we see the odds are 5 to 3.

13. $P(\text{German soldier killed}) = \dfrac{1}{5+1} = \dfrac{1}{6}$

15. The odds against the baby having Down's syndrome are a to b, where

$$\frac{a}{b} = \frac{1 - (1/35)}{1/35} = \frac{34/35}{1/35} = \frac{34}{1}.$$

So we see the odds are 34 to 1.

17. $(2/5)(\$2) = \$0.80$

19. $(10/1)(\$75) = \$750$

21. **(a)** There are 18 ways for the ball to land on red and $38 - 18 = 20$ ways for the ball not to land on red. Therefore, the odds against winning a bet on red are 20 to 18. Dividing by 2, we express these odds as 10 to 9.

    (b) The house odds of 1 to 1 are not equal to the true odds of 10 to 9, so betting on red is not a fair bet.

23. If you believe the horse has a 20% chance of winning, then you believe the true odds are a to b, where

$$\frac{a}{b} = \frac{1 - 0.20}{0.20} = \frac{0.80}{0.20} = \frac{80}{20} = \frac{4}{1}.$$

So, you believe the true odds are 4 to 1. Because the house odds are also 4 to 1, you think wagering on the horse is a fair bet.

SECTION 4.3

1. $P(\text{type B or Rh-negative}) = P(\text{type B}) + P(\text{Rh-negative}) - P(\text{type B and Rh-negative})$
$$= 0.12 + 0.15 - 0.02 = 0.25 = 25\%$$

3. **(a)** $P(\text{children or pets}) = P(\text{children}) + P(\text{pets}) - P(\text{children and pets})$
$$= \frac{35}{60} + \frac{22}{60} - \frac{15}{60} = \frac{42}{60} = \frac{7}{10}$$

    (b) $P(\text{neither children nor pets}) = 1 - P(\text{children and pets}) = 1 - \dfrac{7}{10} = \dfrac{3}{10}$

5. $P(\text{passed reading or math}) = P(\text{passed reading}) + P(\text{passed math}) - P(\text{passed both})$
$$= 0.779 + 0.63 - 0.586 = 0.823 = 82.3\%$$

7. P(approves of neither candidate)

$= 1 - P$(approves of Republican or approves of Democrat)

$= 1 - [P$(approves of Republican) $+ P$(approves of Democrat) $- P$(approves of both)$]$

$= 1 - [0.37 + 0.42 - 0.07] = 1 - 0.72 = 0.28 = 28\%$

9. P(ace or red) $= P$(ace) $+ P$(red) $- P$(ace and red) $= \dfrac{4}{52} + \dfrac{26}{52} - \dfrac{2}{52} = \dfrac{28}{52} = \dfrac{7}{13}$

11. P(over 74 or under 5) $= P$(over 74) $+ P$(under 5)

$$= \frac{13{,}135{,}272}{248{,}709{,}873} + \frac{18{,}354{,}443}{248{,}709{,}873} = \frac{31{,}489{,}715}{248{,}709{,}873} \approx 0.1266$$

13. $P(1, 2, \text{ or } 3) = P(1) + P(2) + P(3) = \dfrac{1}{6} + \dfrac{1}{4} + \dfrac{1}{8} = \dfrac{4}{24} + \dfrac{6}{24} + \dfrac{3}{24} = \dfrac{13}{24}$

15. P(likes hotdogs or pizza) $= P$(likes hotdogs) $+ P$(likes pizza) $- P$(likes hotdogs and pizza)

$$\frac{3}{4} = \frac{1}{3} + \frac{1}{2} - P\text{(likes hotdogs and pizza)}$$

$$P\text{(likes hotdogs and pizza)} = \frac{1}{3} + \frac{1}{2} - \frac{3}{4} = \frac{4}{12} + \frac{6}{12} - \frac{9}{12} = \frac{1}{12}$$

SECTION 4.4

1.   **(a)** P(enrolled) $= \dfrac{6113}{11{,}058} \approx 0.5528$

     (b) P(enrolled | mother had only an elementary education) $= \dfrac{219}{556} \approx 0.3939$

     (c) P(enrolled | mother had BA or higher) $= \dfrac{1476}{2169} \approx 0.6805$

     (d)   Yes, it appears that the higher the education level of the mother the more likely it is that a child of hers is enrolled in preschool or kindergarten.

3.   **(a)** P(male author's paper rejected | editor was male) $= \dfrac{308}{800} = 0.385$

     (b) P(female author's paper rejected | editor was male) $= \dfrac{59}{204} \approx 0.2892$

     (c) P(male author's paper rejected | editor was female) $= \dfrac{278}{605} \approx 0.4595$

     (d) P(female author's paper rejected | editor was female) $= \dfrac{108}{242} \approx 0.4463$

     (e)   Female editors reject papers authored by males at about the same rate as those authored by females, whereas male editors reject papers authored by males at a higher rate than those authored by females. Furthermore, female editors have a higher overall rejection rate than male editors.

5. P(guess correctly) $= 1/3$

7. P(tail and 4) $= P$(tail) $\cdot P(4) = \dfrac{1}{2} \cdot \dfrac{1}{6} = \dfrac{1}{12}$

9. The outcomes of the roll of two dice in which at least one of the dice rolled is a 5 are: (5, 1), (5, 2), (5, 3), (5, 4), (5, 5), (5, 6), (1, 5), (2, 5), (3, 5), (4, 5), (6, 5). Only two of these eleven outcomes give a sum of 9. Therefore,

$$P(\text{sum is } 9 \mid \text{at least one of the dice rolled is a } 5) = \frac{2}{11}.$$

11. $P(\text{both fail}) = (0.0000001)(0.0000001) = 0.00000000000001 = 0.000000000001\%$

13. $P(\text{ace of spades} \mid \text{black card}) = 1/26$

15. (a) $P(\text{all are 5's or 6's}) = \frac{2}{6} \cdot \frac{2}{6} \cdot \frac{2}{6} = \frac{1}{3} \cdot \frac{1}{3} \cdot \frac{1}{3} = \frac{1}{3^3} = \frac{1}{27}$

    (b) $P(\text{none is 5 or 6}) = \frac{4}{6} \cdot \frac{4}{6} \cdot \frac{4}{6} = \frac{2}{3} \cdot \frac{2}{3} \cdot \frac{2}{3} = \frac{2^3}{3^3} = \frac{8}{27}$

17. $P(\text{win the first and second prizes}) = \frac{2}{120} \cdot \frac{1}{119} = \frac{1}{60} \cdot \frac{1}{119} = \frac{1}{7140}$

19. $P(\text{all hearts}) = \frac{13}{52} \cdot \frac{12}{51} \cdot \frac{11}{50} \cdot \frac{10}{49} = \frac{17,160}{6,497,400} \approx 0.002641$

21. $P(\text{college graduate} \mid \text{male}) = \dfrac{P(\text{college graduate and male})}{P(\text{male})} = \dfrac{0.076}{0.571} \approx 0.133$

23. $P(\text{winner}) = \frac{1}{10} \cdot \frac{1}{10} \cdot \frac{1}{10} = \frac{1}{10^3} = \frac{1}{1000}$

25. $P(\text{all grape}) = \frac{7}{25} \cdot \frac{6}{24} \cdot \frac{5}{23} \cdot \frac{4}{22} = \frac{840}{303,600} \approx 0.002767$

27. $P(\text{woman and planning to teach}) = P(\text{woman}) \cdot P(\text{planning to teach} \mid \text{woman}) = (0.56)(0.317) \approx 0.178$

29. (a) $P(\text{all five Caesareans}) = (0.228)(0.228)(0.228)(0.228)(0.228)$
    $$= (0.228)^5 \approx 0.000616$$

    (b) $P(\text{none are Caesareans}) = (1 - 0.228)(1 - 0.228)(1 - 0.228)(1 - 0.228)(1 - 0.228)$
    $$= (0.772)(0.772)(0.772)(0.772)(0.772)$$
    $$= (0.772)^5 = 0.274$$

31. (a) $P(\text{name same tire}) = 1/4$

    (b) $P(\text{name same tire}) = P(\text{both name right front}) + P(\text{both name right rear})$
    $$+ P(\text{both name left front}) + P(\text{both name left rear})$$
    $$= (0.40)(0.40) + (0.30)(0.30) + (0.20)(0.20) + (0.10)(0.10)$$
    $$= 0.16 + 0.09 + 0.04 + 0.01 = 0.3 = 30\%$$

33. $P(\text{found defective}) = P(\text{defective and test finds defective})$
    $$= P(\text{defective}) + P(\text{test finds defective} \mid \text{defective})$$
    $$= (0.02)(0.92) = 0.0184 = 1.84\%$$

35. (a) First note that there were $(0.52)(2200) = 1144$ respondents.

$$P(\text{more than one injury} \mid \text{one or more injuries}) = \frac{P(\text{more than one injury})}{P(\text{one or more injuries})}$$

$$= \frac{(58 + 9)/1144}{0.34} \approx 0.172$$

(b) $P(\text{three injuries} \mid \text{two or more injuries}) = \dfrac{n(\text{three injuries})}{n(\text{two or more injuries})}$

$$= \frac{9}{58 + 9} = \frac{9}{67} \approx 0.134$$

37. $P(\text{not covered} \mid \text{income less than \$25,000}) = \dfrac{P(\text{not covered and income less than \$25,000})}{P(\text{income less than \$25,000})}$

$$= \frac{(18,470,000/265,284,000)}{0.278} \approx 0.250$$

39. First, we compute the probability that a test will be negative.

$$P(\text{negative}) = P(\text{fluid and negative}) + P(\text{no fluid and negative})$$
$$= P(\text{fluid}) \cdot P(\text{negative} \mid \text{fluid}) + P(\text{no fluid}) \cdot P(\text{negative} \mid \text{no fluid})$$
$$= (0.065)(1 - 0.981) + (1 - 0.065)(1 - 0.001)$$
$$= (0.065)(0.019) + (0.935)(0.999)$$
$$= 0.001235 + 0.934065 = 0.9353.$$

We now use this calculation to compute the probability that a patient with a negative test has pericardial fluid.

$$P(\text{fluid} \mid \text{negative}) = \frac{P(\text{fluid and negative})}{P(\text{negative})}$$

$$= \frac{P(\text{fluid}) \cdot P(\text{negative} \mid \text{fluid})}{P(\text{negative})}$$

$$= \frac{(0.065)(1 - 0.981)}{0.9353}$$

$$= \frac{(0.065)(0.019)}{0.9353} \approx 0.00132$$

41. Suppose $P(B|A) < P(B)$. If $P(B|\text{not } A) \leq P(B)$, then

$$P(B) = P(A \text{ and } B) + P(\text{not } A \text{ and } B)$$
$$= P(A) \cdot P(B|A) + P(\text{not } A) \cdot P(B|\text{not } A)$$
$$< P(A) \cdot P(B) + P(\text{not } A) \cdot P(B) = [P(A) + P(\text{not } A)]P(B) = P(B).$$

However, it is not possible that $P(B) < P(B)$, so our assumption that $P(B|\text{not } A) \leq P(B)$ must be false. Therefore, $P(B|\text{not } A) > P(B)$. Similarly, if $P(B|A) > P(B)$, then $P(B|\text{not } A) < P(B)$.

SECTION 4.5

1. $5 \cdot 4 \cdot 5 = 100$

3. $_{52}C_6 = 20{,}358{,}520$

5. (a) $_{12}C_5 = 792$ (b) $_8C_3 \cdot {}_4C_2 = 56 \cdot 6 = 336$ (c) $_8C_5 = 56$

7. $1 \cdot 1 \cdot 8 \cdot 7 \cdot 26 \cdot 25 = 36{,}400$

9. (a) $_{13}C_5 = 1287$ (b) $4 \cdot {}_{13}C_5 = 4 \cdot 1287 = 5148$

11. $_{10}C_2 \cdot {}_8C_2 \cdot {}_{12}C_2 \cdot {}_7C_2 = 45 \cdot 28 \cdot 66 \cdot 21 = 1{,}746{,}360$

13. $2^{15} = 32{,}768$

15. $_4C_2 \cdot {}_{13}C_3 \cdot {}_{13}C_3 = 6 \cdot 286 \cdot 286 = 490{,}776$

17. (a) $36 \cdot 36 \cdot 36 \cdot 36 \cdot 36 \cdot = 36^5 = 60{,}466{,}176$

     (b) $1 \cdot 36 \cdot 36 \cdot 36 \cdot 36 \cdot = 36^4 = 1{,}679{,}616$

19. $6^3 = 216$

21. $10 \cdot 26 \cdot 10 \cdot 10 \cdot 10 \cdot 10 = 2{,}600{,}000$

23. $10 \cdot 10 \cdot 10 \cdot 3 \cdot 2 \cdot 1 = 10^3 \cdot 3! = 6000$

25. $_{435}C_3 = 13{,}624{,}345$

27. $5! = 120$

29. (a) $2^7 = 128$

     (b) A car thief could carry all 128 possible keys and try them until the correct one is found in a short amount of time.

31. $_6P_3 = 120$

33. $13 \cdot {}_4C_3 \cdot 12 \cdot {}_4C_1 = 13 \cdot 4 \cdot 12 \cdot 4 = 2496$

35. $_{15}C_3 \cdot {}_{12}C_3 \cdot {}_9C_3 \cdot {}_6C_3 = 455 \cdot 220 \cdot 84 \cdot 20 = 168{,}168{,}000$

37. $_{49}C_5 \cdot {}_{42}C_1 = 1{,}906{,}884 \cdot 42 = 80{,}089{,}128$

SECTION 4.6

1. $P(\text{winning trifecta bet}) = \dfrac{1}{{}_{11}P_3} = \dfrac{1}{990}$

3. $P(\text{winning}) = \dfrac{1}{{}_{35}C_5} = \dfrac{1}{324{,}632}$

5. $P(\text{correct order}) = \dfrac{1}{10!} = \dfrac{1}{3{,}628{,}800}$

7. $P(\text{open lock}) = \dfrac{1}{{}_{12}C_3} = \dfrac{1}{220}$

9. $P(\text{two hearts and three spades}) = \dfrac{_{13}C_2 \cdot _{13}C_3}{_{52}C_5} = \dfrac{78 \cdot 286}{2{,}598{,}960} = \dfrac{22{,}308}{2{,}598{,}960} \approx 0.008583$

11. **(a)** $P(\text{all correct}) = \dfrac{1}{2^{10}} + \dfrac{1}{1024}$

    (b) $P(\text{exactly 5 correct}) = \dfrac{_{10}C_5}{2^{10}} = \dfrac{252}{1024} \approx 0.2461$

    (c) $P(\text{at least 1 correct}) = 1 - P(\text{all incorrect}) = 1 - \dfrac{1}{1024} = \dfrac{1023}{1024}$

13. $P(\text{matching three winning numbers}) = \dfrac{_4C_3 \cdot _{96}C_1}{_{100}C_4} = \dfrac{4 \cdot 96}{3{,}921{,}225} = \dfrac{384}{3{,}921{,}225} \approx 0.00009793$

15. $P(\text{each child gets own lunch box}) = \dfrac{1}{4!} = \dfrac{1}{24}$

17. $P(\text{one or more repeated digits}) = 1 - P(\text{no repeated digits}) = 1 - \left(\dfrac{8 \cdot _9P_6}{8 \cdot 10^6}\right)$

$$= 1 - \left(\dfrac{483{,}840}{8{,}000{,}000}\right)$$

$$= 1 - 0.06048 \approx 0.9395$$

19. **(a)** $P(\text{all grape}) = \dfrac{_{10}C_3}{_{24}C_3} = \dfrac{120}{2024} \approx 0.05929$

    (b) $P(\text{all same flavor}) = P(\text{all grape}) + P(\text{all orange}) + P(\text{all cherry})$

$$= \dfrac{_{10}C_3}{_{24}C_3} + \dfrac{_6C_3}{_{24}C_3} + \dfrac{_8C_3}{_{24}C_3}$$

$$= \dfrac{120}{2024} + \dfrac{20}{2024} + \dfrac{56}{2024} = \dfrac{196}{2024} \approx 0.09684$$

    (c) $P(\text{one of each flavor}) = \dfrac{_{10}C_1 \cdot _6C_1 \cdot _8C_1}{_{24}C_3} = \dfrac{10 \cdot 6 \cdot 8}{2024} = \dfrac{480}{2024} \approx 0.23715$

21. $P = \dfrac{_{13}C_3 \cdot _4C_2 \cdot _4C_2 \cdot _4C_2}{_{52}C_6} = \dfrac{286 \cdot 6 \cdot 6 \cdot 6}{20{,}358{,}520} = \dfrac{61{,}776}{20{,}358{,}520} \approx 0.003034$

23. $P(\text{all letters the same}) = \dfrac{26 \cdot 1 \cdot 1 \cdot 10^4}{26^3 \cdot 10^4} = \dfrac{1}{26^2} = \dfrac{1}{676} \approx 0.001479$

25. $P(\text{all sweet}) = \dfrac{_{11}C_3}{_{20}C_3} = \dfrac{165}{1140} \approx 0.1447$

27. **(a)** $P(\text{blackjack}) = \dfrac{_4C_1 \cdot _{16}C_1}{_{52}C_2} = \dfrac{4 \cdot 16}{1326} = \dfrac{64}{1326} \approx 0.048265$

    (b) $P(\text{blackjack}) = \dfrac{_8C_1 \cdot _{32}C_1}{_{104}C_2} = \dfrac{8 \cdot 32}{5356} = \dfrac{256}{5356} \approx 0.047797$

29. $P = \dfrac{10 \cdot 10 \cdot 10 \cdot 10}{_{40}P_4} = \dfrac{10{,}000}{2{,}193{,}360} \approx 0.004559$

31. $P(\text{five girls in each class}) = \dfrac{{}_{15}C_5 \cdot {}_{30}C_{10} \cdot {}_{10}C_5 \cdot {}_{20}C_{10}}{{}_{45}C_{15} \cdot {}_{30}C_{15}}$

$$= \frac{3003 \cdot 30{,}045{,}015 \cdot 252 \cdot 184{,}756}{344{,}867{,}425{,}584 \cdot 155{,}117{,}520} \approx 0.07853$$

33. $P(\text{at least two have same birthday}) = 1 - P(\text{all have different birthdays})$

$$= 1 - \frac{{}_{365}P_{15}}{365^{15}} \approx 1 - 0.7471 = 0.2529 = 25.29\%$$

35. $P(\text{at least two have same birth month}) = 1 - P(\text{all have different birth months})$

$$= 1 - \frac{{}_{12}P_7}{12^7} \approx 1 - 0.1114 = 0.8886 = 88.86\%$$

37. $P(\text{winning jackpot}) = \dfrac{1}{{}_{49}C_5 \cdot {}_{42}C_1} = \dfrac{1}{1{,}906{,}884 \cdot 42} = \dfrac{1}{80{,}089{,}128}$

39. $P(\text{matching 2 white balls and Powerball}) = \dfrac{{}_5C_2 \cdot {}_{44}C_3 \cdot 1}{{}_{49}C_5 \cdot {}_{42}C_1}$

$$= \frac{10 \cdot 13{,}244}{1{,}906{,}884 \cdot 42}$$

$$= \frac{132{,}440}{80{,}089{,}128} \approx 0.001654$$

41. $P(\text{matching 3 white balls and missing Powerball}) = \dfrac{{}_5C_3 \cdot {}_{44}C_2 \cdot {}_{41}C_1}{{}_{49}C_5 \cdot {}_{42}C_1} = \dfrac{10 \cdot 946 \cdot 41}{1{,}906{,}884 \cdot 42}$

$$= \frac{387{,}860}{80{,}089{,}128} \approx 0.004843$$

SECTION 4.7

1. expected point value $= P(\text{correct}) \cdot 1 + P(\text{incorrect}) \cdot \left(-\dfrac{1}{4}\right)$

$$= \left(\frac{1}{3}\right) \cdot 1 + \left(\frac{2}{3}\right)\left(-\frac{1}{4}\right) = \frac{1}{3} - \frac{1}{6} = \frac{2}{6} - \frac{1}{6} = \frac{1}{6}$$

3. expected value $= \left(\dfrac{1}{6}\right) \cdot 1 + \left(\dfrac{1}{6}\right) \cdot 2 + \left(\dfrac{1}{6}\right) \cdot 3 + \left(\dfrac{1}{6}\right) \cdot 4 + \left(\dfrac{1}{6}\right) \cdot 5 + \left(\dfrac{1}{6}\right) \cdot 6$

$$= \frac{1}{6} + \frac{2}{6} + \frac{3}{6} + \frac{4}{6} + \frac{5}{6} + \frac{6}{6} = \frac{21}{6} = 3.5$$

5. First we compute the expected payoff and find

$$\text{expected payoff} = P(\text{roll 4 or 5}) \cdot \$6 + P(\text{roll 6}) \cdot \$10$$

$$= \left(\frac{2}{6}\right) \cdot \$6 + \left(\frac{1}{6}\right) \cdot \$10 = \$\left(\frac{12 + 10}{6}\right) = \$\left(\frac{22}{6}\right) \approx \$3.67.$$

Subtracting the \$5 you pay to play we get

$$\text{expected value} \approx \$3.67 - \$5.00 = -\$1.33.$$

7. First we compute the expected payoff, and we get

$$\text{expected payoff} = \left(\frac{1}{2000}\right) \cdot \$1000 + \left(\frac{3}{2000}\right) \cdot \$100 + \left(\frac{5}{2000}\right) \cdot \$20 + \left(\frac{10}{2000}\right) \cdot \$5$$

$$= \$\left(\frac{1000 + 300 + 100 + 50}{2000}\right) = \$\left(\frac{1450}{2000}\right) = \$0.725.$$

Subtracting the $1 fee we find

$$\text{expected value} = \$0.725 - \$1.00 = -\$0.275.$$

9. The expected payoff of a dart throw is given by

$$\text{expected payoff} = (0.15) \cdot \$5 + (0.05) \cdot \$50 = \$3.25.$$

So, to make an average profit of 50¢ per player, the carnival organizers should charge

$$\$3.25 + \$0.50 = \$3.75.$$

11. (a) $\text{expected value} = P(\text{winning}) \cdot \$1 + P(\text{losing}) \cdot (-\$1)$

$$= \left(\frac{18}{38}\right) \cdot \$1 + \left(\frac{20}{38}\right) \cdot (-\$1) = \$\left(\frac{18 - 20}{38}\right)$$

$$= -\$\left(\frac{2}{38}\right) \approx -\$0.0526$$

(b) 5.26%

13. (a) The expected payoff on a card is given by

$$\text{expected payoff} = P(\text{winning}) \cdot \$500 = \left(\frac{1}{2500}\right) \cdot \$500 = \$0.20.$$

Subtracting the 25¢ cost of the card we find

$$\text{expected value} = \$0.20 - \$0.25 = -\$0.05.$$

Therefore, your expected loss is $0.05 or 5¢.

(b) The expected payoff on a card is given by

$$\text{expected payoff} = P(\text{winning}) \cdot \$500 = \left(\frac{1}{800}\right) \cdot \$500 = \$0.625.$$

Subtracting the 25¢ cost of the card we find

$$\text{expected value} = \$0.625 - \$0.25 = \$0.375.$$

Therefore, your expected gain is $0.375 or 37.5¢.

15. (a) First we compute the expected payoff and find

$$\text{expected payoff} = P(\text{three winning numbers}) \cdot \$2 + P(\text{four winning numbers}) \cdot \$20$$

$$+ P(\text{five winning numbers}) \cdot \$480$$

$$= \left(\frac{20C_3 \cdot 60C_2}{80C_5}\right) \cdot \$2 + \left(\frac{20C_4 \cdot 60C_1}{80C_5}\right) \cdot \$20 + \left(\frac{20C_5}{80C_5}\right) \cdot \$480$$

$$\approx (0.083935) \cdot \$2 + (0.012092) \cdot \$20 + (0.000645) \cdot \$480$$

$$\approx \$0.719.$$

Subtracting the $1 cost of playing, we find

$$\text{expected value} \approx \$0.719 - \$1.00 = -\$0.281.$$

(b) 28.1%

17. (a) $\text{expected value} = \left(\dfrac{1}{100{,}000}\right) \cdot \$9360 + \left(\dfrac{5}{100{,}000}\right) \cdot \$2000 + \left(\dfrac{50}{100{,}000}\right) \cdot \$156$

$$+ \left(\dfrac{500}{100{,}000}\right) \cdot \$16 + \left(\dfrac{5000}{100{,}000}\right) \cdot \$3 + \left(\dfrac{20{,}000}{100{,}000}\right) \cdot \$0.50$$

$$= \$\left(\dfrac{9360 + 10{,}000 + 7800 + 8000 + 15{,}000 + 10{,}000}{100{,}000}\right)$$

$$= \$\left(\dfrac{60{,}160}{100{,}000}\right) = \$0.6016$$

(b) $\text{expected value} = \$0.6016 - \$0.40 = \$0.2016$

(c) $\text{expected value} = \$0.2016 - \left(\dfrac{20{,}000}{100{,}000}\right) \cdot \$0.50 = \$0.2016 - \$0.1 = \$0.1016$

19. (a)

Number of Games	Expected Value
20	$(1/2^{20}) \cdot \$100{,}000 \approx \$0.095$
16	$(1/2^{16}) \cdot \$10{,}000 \approx \$0.153$
12	$(1/2^{12}) \cdot \$500 \approx \$0.122$
10	$(1/2^{10}) \cdot \$50 \approx \$0.049$
8	$(1/2^{8}) \cdot \$15 \approx \$0.059$
5	$(1/2^{5}) \cdot \$5 \approx \$0.156$

(b) 5

(c) Because all of the expected values computed in (a) are less than 40¢, the expected value is never positive if the 40¢ cost of a stamp and envelope is taken into account.

21. $\text{expected number} = P(\text{rain}) \cdot 200 + P(\text{no rain}) \cdot 300$

$$= \left(\dfrac{115}{365}\right) \cdot 200 + \left(1 - \dfrac{115}{365}\right) \cdot 300 = \left(\dfrac{115}{365}\right) \cdot 200 + \left(\dfrac{250}{365}\right) \cdot 300$$

$$= \dfrac{23{,}000 + 75{,}000}{365} = \dfrac{98{,}000}{365} \approx 268.49 \text{ customers}$$

23. (a) The expected value of the amount the man would have to pay the first lawyer is simply the $12,000 flat fee. The expected value of the amount the man would have to pay the second lawyer is given by

$$\text{expected value} = P(\text{winning}) \cdot \$30{,}000 = (0.75) \cdot \$30{,}000 = \$22{,}500.$$

Therefore, the man should choose the first lawyer.

(b) If the man did not have $12,000 available to pay the first lawyer if he lost the lawsuit, then he might choose the second lawyer who would only have to be paid if the lawsuit is successful.

SECTION 4.8

1. (a)

	T	s
s	Ts	ss
s	Ts	ss

    (b) $P(\text{tall}) = 1/2$

    (c) $P(\text{short}) = 1/2$

3. The Punnett square is

	b	b
b	bb	bb
b	bb	bb

    so $P(\text{blue eyes}) = 1$.

5. (a) Both the husband and wife must have gene pair Bb, so the Punnett square is

	B	b
B	BB	Bb
b	Bb	bb

    and therefore $P(\text{brown eyes}) = 3/4$.

    (b) Because the second child has brown eyes, we see from the Punnett square in (a) that the second child has gene pair BB with probability 1/3 and gene pair Bb with probability 2/3. If the second child has gene pair BB, then the Punnett square for the child of the second child and the blue-eyed spouse is:

	B	B
b	Bb	Bb
b	Bb	Bb

    If the gene type of the second child is Bb, then the Punnett square for the child of the second child and the blue-eyed spouse is

	B	b
b	Bb	bb
b	Bb	bb

    Therefore, for the child of the second child and the blue-eyed spouse we have

    $$P(\text{brown eyes}) = P(\text{BB parent and brown eyes}) + P(\text{Bb parent and brown eyes})$$

    $$= P(\text{BB parent}) \cdot P(\text{brown eyes} \,|\, \text{BB parent})$$

    $$+ P(\text{Bb parent}) \cdot P(\text{brown eyes} \,|\, \text{Bb parent})$$

    $$= \left(\frac{1}{3} \cdot 1\right) + \left(\frac{2}{3} \cdot \frac{1}{2}\right) = \frac{1}{3} + \frac{1}{3} = \frac{2}{3}.$$

7. The Punnett square is

	T	t
T	TT	Tt
t	Tt	tt

so $P(\text{carrier} \,|\, \text{does not have Tay-Sachs}) = 2/3$.

9. The Punnett square for the children is

	H	h
h	Hh	hh
h	Hh	hh

(a) $P(\text{all four develop disease}) = \dfrac{1}{2} \cdot \dfrac{1}{2} \cdot \dfrac{1}{2} \cdot \dfrac{1}{2} = \dfrac{1}{16}$

(b) $P(\text{at least one develops disease}) = 1 - P(\text{none develop disease})$

$$= 1 - \left(\dfrac{1}{2} \cdot \dfrac{1}{2} \cdot \dfrac{1}{2} \cdot \dfrac{1}{2}\right) = 1 - \dfrac{1}{16} = \dfrac{15}{16}$$

11. The Punnett square is

	R	R
R	RR	RR
W	RW	RW

(a) $P(\text{red}) = 1/2$ (b) $P(\text{white}) = 0$ (c) $P(\text{pink}) = 1/2$

13. The Punnett square is

	R	W
R	RR	RW
W	RW	WW

(a) $P(\text{red}) = 1/4$ (b) $P(\text{white}) = 1/4$ (c) $P(\text{pink}) = 1/2$

15. The Punnett square is

	X	X_h
X	XX	XX_h
Y	XY	$X_h Y$

so $P(\text{carrier} \,|\, \text{female}) = 1/2$.

17. The Punnett square is

	X	X_c
X_c	XX_c	$X_c X_c$
Y	XY	$X_c Y$

(a) $P(\text{colorblind}) = 1/2$ (b) $P(\text{carrier}) = 1/4$

19. The Punnett square is

	X	X
X_R	$X_R X$	$X_R X$
Y	XY	XY

so $P(\text{disease}) = 1/2$.

21. The Punnett square is

	A	B
o	Ao	Bo
o	Ao	Bo

(a) $P(A) = 1/2$ (b) $P(B) = 1/2$

(c) $P(O) = 0$ (d) $P(AB) = 0$

23. Because the man's father was type A, the man must have gene pair Bo. Therefore, the Punnett square for the man and his wife is

	B	o
o	Bo	oo
o	Bo	oo

(a) $P(A) = 0$ (b) $P(B) = 1/2$

(c) $P(O) = 1/2$ (d) $P(AB) = 0$

25. Because the couple has a child with Rh-negative, type B blood, the wife must have gene pairs Ao and Rr. Therefore, the Punnett square for the blood type of the baby is

	A	o
A	AA	Ao
B	AB	Bo

and the Punnett square for the Rh factor of the baby is

	R	r
r	Rr	rr
r	Rr	rr

Therefore, for the baby,

$$P(\text{Rh-negative and type A}) = P(\text{Rh-negative}) \cdot P(\text{type A}) = \frac{1}{2} \cdot \frac{1}{2} = \frac{1}{4}.$$

27. Because neither of Mr. Rosen's parents have the disease and his sister has the disease, the Punnett square for Mr. Rosen is

	D	d
D	DD	Dd
d	Dd	dd

where d denotes the disease gene and D the normal gene. Mr. Rosen can pass the disease on to his child only if he is a carrier. In this case, the Punnett square for Mr. and Mrs. Rosen's baby would be

	D	d
d	Dd	dd
d	Dd	dd

Therefore,

$P(\text{child inherits disease}) = P(\text{Mr. Rosen a carrier and child inherits disease})$

$= P(\text{Mr. Rosen a carrier}) \cdot P(\text{child inherits disease} \mid \text{Mr. Rosen a carrier})$

$= \dfrac{2}{3} \cdot \dfrac{1}{2} = \dfrac{1}{3}.$

Chapter 4 Review Exercises

1. The outcomes that have a sum of 10 are (6, 4), (4, 6), and (5, 5). There are 36 possible outcomes. Therefore, $P(\text{sum of }10) = 3/36 = 1/12$.

3. $P(\text{pop}) = \dfrac{253}{300} \approx 0.8433$

5. The odds against winning are a to b, where

$$\frac{a}{b} = \frac{1 - (1/3)}{1/3} = \frac{2/3}{1/3} = \frac{2}{1}.$$

So we see the odds are 2 to 1.

7. The true odds against winning the game are a to b, where

$$\frac{a}{b} = \frac{1 - 0.15}{0.15} = \frac{0.85}{0.15} = \frac{85}{15} = \frac{17}{3}.$$

So we see the true odds are 17 to 3. Because the house odds of 19 to 4 are not equal to the true odds of 17 to 3, we see that the bet is not a fair bet.

9. $P(\text{neither woman nor commissioned officer})$

$= 1 - P(\text{woman or commissioned officer})$

$= 1 - [P(\text{woman}) + P(\text{commissioned officer}) - P(\text{woman and commissioned officer})]$

$= 1 - \left[\dfrac{70{,}690}{487{,}297} + \dfrac{67{,}986}{487{,}297} - \dfrac{9716}{487{,}297} \right]$

$= 1 - \left[\dfrac{128{,}960}{487{,}297} \right] = \dfrac{358{,}337}{487{,}297} \approx 0.7354$

11. $P(\text{both diamonds}) = \dfrac{13}{52} \cdot \dfrac{12}{51} = \dfrac{156}{2652} \approx 0.05882$

13. $P(\text{under 15} \mid \text{urgent}) = \dfrac{P(\text{under 15 and urgent})}{P(\text{urgent})} = \dfrac{P(\text{under 15}) \cdot P(\text{urgent} \mid \text{under 15})}{P(\text{urgent})}$

$= \dfrac{(0.251)(0.394)}{(0.453)} \approx 0.218$

15. $_7C_2 = 21$

17. $P(\text{four women and four men}) = \dfrac{_{14}C_4 \cdot {}_{11}C_4}{_{25}C_8} = \dfrac{1001 \cdot 330}{1{,}081{,}575} = \dfrac{330{,}330}{1{,}081{,}575} \approx 0.3054$

19. $P(\text{three heads}) = \dfrac{{}_7C_3}{2^7} = \dfrac{35}{128} \approx 0.2734$

21. $P = \dfrac{4 \cdot {}_{13}C_4 \cdot 3 \cdot {}_{13}C_2 \cdot 2 \cdot {}_{13}C_1}{{}_{52}C_7} = \dfrac{4 \cdot 715 \cdot 3 \cdot 78 \cdot 2 \cdot 13}{133{,}784{,}560} = \dfrac{17{,}400{,}240}{133{,}784{,}560} \approx 0.1301$

23. For the first choice the expected change in value of the investment is simply the guaranteed return of $(0.05)(\$100{,}000) = \$5000$, so the expected value of the investment is $\$105{,}000$. For the second choice the expected change in value of the investment is given by

$$\text{expected gain} = P(\text{winning bet}) \cdot \$100{,}000 + P(\text{losing bet}) \cdot (-\$100{,}000)$$
$$= (0.51) \cdot \$100{,}000 + (0.49) \cdot (-\$100{,}000)$$
$$= \$2000,$$

so the expected value of the investment is $\$102{,}000$. Therefore, the first choice has the highest expected value.

25. Both the husband and wife must have gene pair Bb, so the Punnett square for the children is

	B	b
B	BB	Bb
b	Bb	bb

   (a) $P(\text{blue eyes}) = 1/4$

   (b) $P(\text{at least one will have blue eyes}) = 1 - P(\text{none have blue eyes})$

$$= 1 - \left(\frac{3}{4} \cdot \frac{3}{4} \cdot \frac{3}{4} \cdot \frac{3}{4}\right) = 1 - \frac{81}{256} = \frac{175}{256}$$

27. The Punnett square is

	X	X_c
X	XX	XX_c
Y	XY	X_cY

   (a) $P(\text{colorblind}) = 1/4$

   (b) $P(\text{carrier}) = 1/4$

CHAPTER 5
Student Solution Manual

SECTION 5.1

1. **(a)** The total number of students is 14.

Grade	Frequency	Relative Frequency
A	3	0.2143
B	3	0.2143
C	5	0.3571
D	2	0.1429
F	1	0.0714

(b)

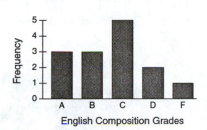

(c)

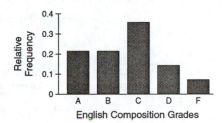

(d) We first compute the angle measures by multiplying relative frequencies by 360°.

Grade	Angle Measure (degrees)
A	77.1
B	77.1
C	128.6
D	51.4
F	25.7

English Composition Grades

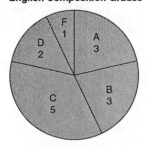

Total = 14 students

3. **(a)**

Number of Justices	Frequency	Relative Frequency
5	19	0.38
7	26	0.52
9	5	0.10

(b)

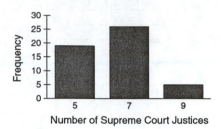

(c)

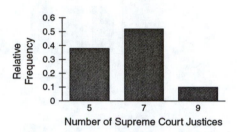

(d)

Number of Justices	Angle Measure (degrees)
5	136.8
7	187.2
9	36.0

Number of Supreme Court Justices

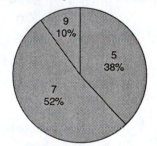

Total = 50 states

(e) All the data are odd numbers in order to avoid tie votes.

5. **(a)** The total world production is 77,920 billion cubic feet.

Region	Relative Frequency
North America	0.3231
Central and South America	0.0331
Western Europe	0.1124
Eastern Europe	0.3327
Middle East	0.0640
Africa	0.0386
Far East and Oceania	0.0961

(b)

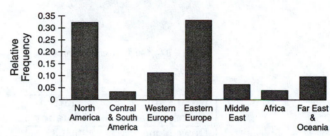

Natural Gas Production

7. (a) The total number of livestock is 874,190.

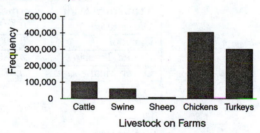

Livestock on Farms

(b)

Type	Relative Frequency
Cattle	0.1161
Swine	0.0685
Sheep	0.0091
Chickens	0.4616
Turkeys	0.3448

→ . X 360°

(c)

Type	Angle Measure (degrees)
Cattle	41.8
Swine	24.7
Sheep	3.3
Chickens	166.2
Turkeys	124.1

Livestock on Farms

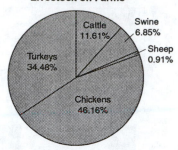

Total = 874,190 thousand animals

9. (a)

Means of Transportation	Relative Frequency
Drove alone	0.803
Carpool	0.148
Public Transit	0.019
Other Means	0.030

(b)

Means of Transportation	Relative Frequency
Drove alone	0.535
Carpool	0.105
Public Transit	0.285
Other Means	0.075

(c)

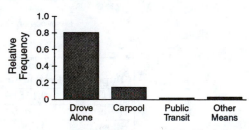

Means of Transportation in Charlotte

(d)

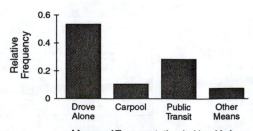

Means of Transportation in New York

11. (a)

Family Type	Angle Measure (degrees)
Two parents	306.0
Mother only	39.6
Father only	3.6
No parent	10.8

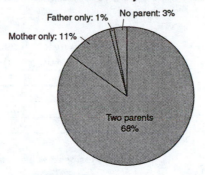

Distribution of Family Structure in 1970

(b)

Family Type	Angle Measure (degrees)
Two parents	244.8
Mother only	86.4
Father only	14.4
No parent	14.4

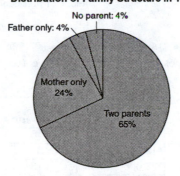

Distribution of Family Structure in 1996

(c) There was a significantly lower percentage of two-parent families in 1996 than in 1970.

13. (a) The youngest age of the 41 presidents is 42, the oldest 69, a difference of 27 years. Our intervals need width at least $27/4 = 6.75$. Intervals of 8 numbers starting at 40 work nicely.

Age	Frequency	Relative Frequency
40-47	5	0.1220
48-55	18	0.4390
56-63	13	0.3171
64-71	5	0.1220

(b)

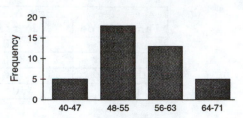

Age of Presidents at Inauguration

(c)

Age	Angle Measure (degrees)
40-47	43.9
48-55	158.0
56-63	114.2
64-71	43.9

Age of Presidents at Inauguration

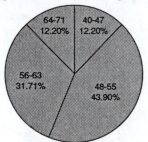

Total = 41 Presidents

15. (a) The smallest number is $15,839 and the largest is $34,454, a difference of $18,615. Noting that 18,615/5 = 3723, one possible grouping and the corresponding frequency and relative frequency distribution is:

Sales (in dollars)	Frequency	Relative Frequency
15,000-18,999	6	0.1176
19,000-22,999	30	0.5882
23,000-26,999	13	0.2549
27,000-30,999	1	0.0196
31,000-34,999	1	0.0196

(b)

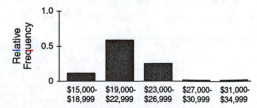

Retail Sales per Household by State, 1993

(c)

Sales (in dollars)	Angle Measure (degrees)
15,000-18,999	42.3
19,000-22,999	211.8
23,000-26,999	91.8
27,000-30,999	7.1
31,000-34,999	7.1

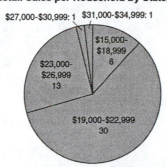

Retail Sales per Household by State, 1993

Total = 50 states and D.C.

17. We begin by computing the relative frequencies for both groupings.

Heights	Frequency	Relative Frequency	Relative Frequency
66.5-67.5	2	0.1250	
67.5-68.5	2	0.1250	0.2500
68.5-69.5	2	0.1250	
69.5-70.5	3	0.1875	0.3125
70.5-71.5	3	0.1875	
71.5-72.5	4	0.2500	0.4375

(a)

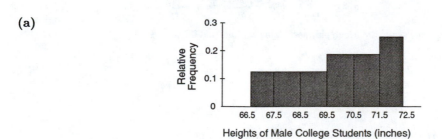

Heights of Male College Students (inches)

(b)

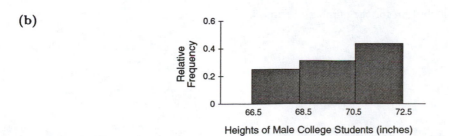

Heights of Male College Students (inches)

19. We begin by finding the frequency distribution that follows.

Tube Size	Frequency
2.75-3.75	2
3.75-4.75	2
4.75-5.75	1
5.75-6.75	22
6.75-7.75	1

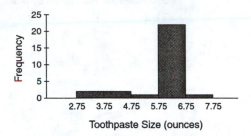

21. The smallest rate is 2.2 and the largest is 8.3. A good starting point is 1.95.

Unemployment Rate	Frequency	Relative Frequency
1.95-3.95	4	0.0784
3.95-5.95	34	0.6667
5.95-7.95	12	0.2353
7.95-9.95	1	0.0196

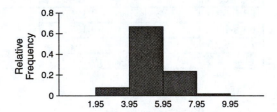

SECTION 5.2

1. There are 3 A's, 3 B's, 5 C's, 2 D's, and 1 F, so the mode is C.

3. The frequency distribution for the number of wins is:

Number of Wins	Frequency
1	1
3	2
4	1
5	1
6	3
7	4
8	4
9	5
10	4
11	1
12	2
13	2

(a) The mode is 9 wins.

(b) median $= \dfrac{8+8}{2} = 8$ wins

(c) $\mu = \dfrac{1\cdot 1 + 2\cdot 3 + 1\cdot 4 + 1\cdot 5 + 3\cdot 6 + 4\cdot 7 + 4\cdot 8 + 5\cdot 9 + 4\cdot 10 + 1\cdot 11 + 2\cdot 12 + 2\cdot 13}{1+2+1+1+3+4+4+5+4+1+2+2}$

$= \dfrac{240}{30} = 8$ wins

5. (a) median $= \dfrac{58.6 + 63.3}{2} = \dfrac{121.9}{2} = 60.95°F$

(b) $\mu = \dfrac{29.0 + 33.5 + 45.8 + 58.6 + 70.1 + 79.6 + 83.7 + 81.8 + 74.8 + 63.3 + 48.4 + 34.0}{12}$

$= \dfrac{702.6}{12} = 58.55°F$

7. (a) median $= 4.3\%$

(b) $\mu = \dfrac{5.2 + 4.3 + 4.1 + 4.8 + 4.5 + 4.0 + 4.0 + 4.3 + 4.2 + 5.0 + 5.4}{11}$

$= \dfrac{49.8}{11} \approx 4.53\%$

9. (a) median $= \$105,250$

(b) $\mu = \dfrac{123,900 + 95,200 + 98,620 + 210,000 + 105,250}{5} = \dfrac{632,970}{5} = \$126,594$

(c) The mean is greatly influenced by the one high sales price, so the median seems more representative.

11. (a) median $= \dfrac{66.3 + 67.3}{2} = 66.8$

(b) $\mu = \dfrac{65.7 + 65.5 + 66.3 + 65.4 + 65.8 + 67.3 + 69.2 + 70.9 + 69.6 + 68.9}{10}$

$= \dfrac{674.6}{10} = 67.46$

(c) The birth rate seems to be increasing.

13. (a) $\mu = \dfrac{52.6 + 39.8 + 53.1 + 36.4 + 50.1 + 36.5 + 55.1 + 38.8 + 49.1}{9} = \dfrac{411.5}{9} \approx 45.7\%$

(b) $\mu = \dfrac{52.6 + 53.1 + 50.1 + 55.1 + 49.1}{5} = \dfrac{260}{5} = 52.0\%$

(c) $\mu = \dfrac{39.8 + 36.4 + 36.5 + 38.8}{4} = \dfrac{151.5}{4} \approx 37.9\%$

(d) The years in part (b) are presidential election years whereas those in part (c) are not.

15. (a) It is bimodal with modes 10¢ and 11¢.

(b) $\text{median} = \dfrac{12 + 13}{2} = 12.5¢$

(c) $\mu = \dfrac{4 \cdot 10 + 4 \cdot 11 + 1 \cdot 12 + 1 \cdot 13 + 3 \cdot 14 + 2 \cdot 16 + 1 \cdot 22 + 1 \cdot 23 + 1 \cdot 25}{4 + 4 + 1 + 1 + 3 + 2 + 1 + 1 + 1}$

$= \dfrac{253}{18} \approx 14.06¢$

17. (a) mode = 16 years old

(b) median = 16 years old

(c) $\mu = \dfrac{3 \cdot 15 + 28 \cdot 16 + 12 \cdot 17 + 8 \cdot 18}{3 + 28 + 12 + 8} = \dfrac{841}{51} \approx 16.5$ years old

19. (a) mode = 18 years old

(b) $\text{median} = \dfrac{17 + 17}{2} = 17$ years old

(c) $\mu = \dfrac{2 \cdot 14 + 10 \cdot 16 + 4 \cdot 17 + 14 \cdot 18}{2 + 10 + 4 + 14} = \dfrac{508}{30} \approx 16.9$ years old

21. (a) $\mu \approx (0 \cdot 1750 + 105 \cdot 2000 + 375 \cdot 2250 + 406 \cdot 2500 + 281 \cdot 2750 + 828 \cdot 3062.5$

$\qquad + 1029 \cdot 3500 + 1089 \cdot 4000 + 1791 \cdot 4500 + 1505 \cdot 5000 + 828 \cdot 5500)/$

$\qquad (0 + 105 + 375 + 406 + 281 + 828 + 1029 + 1089 + 1791 + 1505 + 828)$

$\qquad = 33{,}473{,}250/8237 \approx 4064$ lbs

(b) $\mu \approx (1 \cdot 1750 + 109 \cdot 2000 + 107 \cdot 2250 + 1183 \cdot 2500 + 999 \cdot 2750 + 3071 \cdot 3062.5$

$\qquad + 2877 \cdot 3500 + 1217 \cdot 4000 + 71 \cdot 4500 + 0 \cdot 5000 + 1 \cdot 5500)/$

$\qquad (1 + 109 + 107 + 1183 + 999 + 3071 + 2877 + 1217 + 71 + 0 + 1)$

$\qquad = (30{,}832{,}687.5)/9636 \approx 3200$ lbs

(c) From 1975 to 1990 weights decreased significantly.

23. (a) $\mu \approx (866 \cdot 6 + 3069 \cdot 16 + 3559 \cdot 22 + 2814 \cdot 27 + 2526 \cdot 32 + 1966 \cdot 37 + 1517 \cdot 42 + 993 \cdot 47$

$\qquad + 645 \cdot 52 + 1337 \cdot 64.5 + 414 \cdot 75)/$

$\qquad (866 + 3069 + 3559 + 2814 + 2526 + 1966 + 1517 + 993 + 645 + 1337 + 414)$

$\qquad = 623{,}361.5/19{,}706 \approx 31.6$ years old

(b) $\mu \approx (866 \cdot 6 + 3069 \cdot 16 + 3559 \cdot 22 + 2814 \cdot 27 + 2526 \cdot 32 + 1966 \cdot 37 + 1517 \cdot 42 + 993 \cdot 47$

$\qquad + 645 \cdot 52 + 1337 \cdot 64.5 + 414 \cdot 80)/$

$\qquad (866 + 3069 + 3559 + 2814 + 2526 + 1966 + 1517 + 993 + 645 + 1337 + 414)$

$\qquad = 625{,}431.5/19{,}706 \approx 31.7$ years old

(c) $\mu \approx (866 \cdot 6 + 3069 \cdot 16 + 3559 \cdot 22 + 2814 \cdot 27 + 2526 \cdot 32 + 1966 \cdot 37 + 1517 \cdot 42 + 993 \cdot 47$

$+ 645 \cdot 52 + 1337 \cdot 64.5 + 414 \cdot 85)/$

$(866 + 3069 + 3559 + 2814 + 2526 + 1966 + 1517 + 993 + 645 + 1337 + 414)$

$= 627{,}501.5/19{,}706 \approx 31.8$ years old

(d)   Because 80 seems a reasonable guess of the average age of those 75 and over, we would choose 80 as the representative value.

(e)   The estimates we found in parts (a) through (c) vary only slightly, so having an accurate representative for the "75 and over" category is not overly critical.

25.   A simple example is a class of 10 students where 9 receive a 100% on the exam and 1 receives a 0%. In this case the mean score is given by

$$\mu = \frac{100 + 100 + 100 + 100 + 100 + 100 + 100 + 100 + 100 + 0}{10} = \frac{900}{10} = 90,$$

and we see that 90% of the students have an above average score.

SECTION 5.3

1.   (a) $\mu = \dfrac{25 + 170 + 104 + 413 + 333}{5} = \dfrac{1045}{5} = 209$ sq. mi.

(b)   range $= 413 - 25 = 388$ sq. mi.

(c) $\sigma = \sqrt{\dfrac{(25 - \mu)^2 + (170 - \mu)^2 + (104 - \mu)^2 + (413 - \mu)^2 + (333 - \mu)^2}{5}} \approx 143.8$ sq. mi.

3.   (a) $\mu = \dfrac{4 + 5 + 4 + 7 + 8 + 7 + 6 - 3 - 12 + 1}{10} = \dfrac{27}{10} = \$2.7$ billion

(b)   range $= 8 - (-12) = \$20$ billion

(c) $\sigma = \sqrt{\dfrac{(4-\mu)^2 + (5-\mu)^2 + (4-\mu)^2 + (7-\mu)^2 + (8-\mu)^2 + (7-\mu)^2 + (6-\mu)^2 + (-3-\mu)^2 + (-12-\mu)^2 + (1-\mu)^2}{10}}$

$\approx \$5.8$ billion

5.   (a) $\mu = \dfrac{252 + 195 + 225 + 190 + 225 + 255 + 301 + 256 + 185 + 215 + 225 + 175}{12}$

$= \dfrac{2699}{12} \approx 224.9$ lbs

(b)   range $= 301 - 175 = 126$ lbs

(c) $\mu = \sqrt{\dfrac{(252 - \mu)^2 + (195 - \mu)^2 + \cdots + (175 - \mu)^2}{12}} \approx 35.0$ lbs

7.   (a)   Writing the heights in inches we find

$$\mu = \frac{78 + 79 + 80 + 76 + 79 + 84 + 85 + 81 + 79 + 77 + 85 + 73}{12}$$

$$= \frac{956}{12} \approx 79.67'' = 6'\, 7.67''$$

(b) range $= 85 - 73 = 12''$

(c) $\mu = \sqrt{\dfrac{(78-\mu)^2 + (79-\mu)^2 + \cdots + (73-\mu)^2}{12}} \approx 3.50''$

(d) The difference between the mean height of the Dream Team and the mean height of American males is given by $79.67'' - 70'' = 9.67''$, so we see that the mean height of the Dream Team is well above the national mean.

9. **(a)** $\mu = \dfrac{7.3 + 9.2 + 8.1 + 9.2 + 8.5 + 8.6 + 8.5}{7} = \dfrac{59.4}{7} \approx 8.49\%$

(b) range $= 9.2 - 7.3 = 1.9\%$

(c) $\sigma = \sqrt{\dfrac{(7.3-\mu)^2 + (9.2-\mu)^2 + (8.1-\mu)^2 + (9.2-\mu)^2 + (8.5-\mu)^2 + (8.6-\mu)^2 + (8.5-\mu)^2}{7}}$

$\approx 0.61\%$

(d) The range and standard deviation are fairly small relative to the mean. Therefore, the calculations suggest that the tax burden did not vary much over this period.

11. **(a)** $\mu = \dfrac{5.38 + 3.99 + 3.54 + 2.33 + 1.70 + 1.50 + 0.76 + 1.14 + 1.88 + 3.23 + 5.83 + 5.91}{12}$

$= \dfrac{37.19}{12} \approx 3.10$ inches

(b) $\mu = \dfrac{2.54 + 2.39 + 3.41 + 3.15 + 3.59 + 3.71 + 3.75 + 3.21 + 2.97 + 2.36 + 2.85 + 2.92}{12}$

$= \dfrac{36.85}{12} \approx 3.07$ inches

(c) range $= 5.91 - 0.76 = 5.15$ inches

$$\sigma = \sqrt{\dfrac{(5.83-\mu)^2 + (3.99-\mu)^2 + \cdots + (5.91-\mu)^2}{12}} \approx 1.77 \text{ inches}$$

(d) range $= 3.75 - 2.36 = 1.39$ inches

$$\sigma = \sqrt{\dfrac{(2.54-\mu)^2 + (2.39-\mu)^2 + \cdots + (2.92-\mu)^2}{12}} \approx 0.47 \text{ inches}$$

(e) The monthly precipitation is more variable (seasonal) in Seattle than in Pittsburgh.

13. **(a)** median $= 89.0$ sq. mi.

(b) $\mu = \dfrac{211.3 + 383.3 + 177.3 + 35.8 + 174.3 + 89.0 + 66.6 + 17.8 + 46.5}{9}$

$= \dfrac{1201.9}{9} \approx 133.54$ sq. mi.

(c) The low density states have the largest areas, whereas high-density states have the smallest areas.

(d) range $= 383.3 - 17.8 = 365.5$ sq. mi.

(e) $\sigma = \sqrt{\dfrac{(211.3-\mu)^2 + (383.3-\mu)^2 + (177.3-\mu)^2 + (35.8-\mu)^2 + (174.3-\mu)^2 + (89.0-\mu)^2 + (66.6-\mu)^2 + (17.8-\mu)^2 + (46.5-\mu)^2}{9}}$

≈ 110.05 sq. mi.

15. **(a)** $\mu = \dfrac{2 \cdot 22 + 1 \cdot 23 + 1 \cdot 25 + 2 \cdot 26 + 1 \cdot 27 + 2 \cdot 28 + 1 \cdot 29 + 1 \cdot 30 + 1 \cdot 34}{2 + 1 + 1 + 2 + 1 + 2 + 1 + 1 + 1}$

$= \dfrac{320}{12} \approx 26.67$

(b) range $= 34 - 22 = 12$

(c) $\sigma = \sqrt{\dfrac{2 \cdot (22-\mu)^2 + 1 \cdot (23-\mu)^2 + 1 \cdot (25-\mu)^2 + 2 \cdot (26-\mu)^2 + 1 \cdot (27-\mu)^2 + 2 \cdot (28-\mu)^2 + 1 \cdot (29-\mu)^2 + 1 \cdot (30-\mu)^2 + 1 \cdot (34-\mu)^2}{12}}$

≈ 3.35

17. $\mu = \dfrac{3 \cdot 8.5 + 6 \cdot 8.6 + 3 \cdot 8.7 + 7 \cdot 8.8 + 1 \cdot 8.9}{3 + 6 + 3 + 7 + 1} = \dfrac{173.7}{20} \approx 8.69$

$\sigma = \sqrt{\dfrac{3 \cdot (8.5 - \mu)^2 + 6 \cdot (8.6 - \mu)^2 + 3 \cdot (8.7 - \mu)^2 + 7 \cdot (8.8 - \mu)^2 + 1 \cdot (8.9 - \mu)^2}{20}} \approx 0.12$

19. $\mu = \dfrac{2 \cdot 4999.5 + 18 \cdot 6999.5 + 21 \cdot 8999.5 + 21 \cdot 10{,}999.5 + 2 \cdot 12{,}999.5 + 2 \cdot 14{,}999.5 + 3 \cdot 16{,}999.5 + 1 \cdot 19{,}0000}{2 + 18 + 21 + 21 + 2 + 2 + 3 + 1}$

$= \dfrac{681{,}965.5}{70} \approx 9742$

$\sigma = \sqrt{\dfrac{2 \cdot (4999.5 - \mu)^2 + 18 \cdot (6999.5 - \mu)^2 + \cdots + 1 \cdot (19{,}000 - \mu)^2}{70}} \approx 2791$

21. Using 70 as the representative for the "60 and over" category we arrive at the following values:

$\mu = \dfrac{1271 \cdot 2.5 + 4992 \cdot 9.5 + 15{,}115 \cdot 22 + 9473 \cdot 32 + 8684 \cdot 37 + 40{,}587 \cdot 40 + 14{,}075 \cdot 44.5 + 13{,}366 \cdot 54 + 9878 \cdot 70}{1271 + 4992 + 15{,}115 + 9473 + 8684 + 40{,}587 + 14{,}075 + 13{,}366 + 9878}$

$= \dfrac{4{,}670{,}617}{117{,}441} \approx 39.8 \text{ hours}$

$\sigma = \sqrt{\dfrac{1271 \cdot (2.5 - \mu)^2 + 4992 \cdot (9.5 - \mu)^2 + \cdots + 9878 \cdot (70 - \mu)^2}{117{,}441}} \approx 14.2 \text{ hours}$

SECTION 5.4

1. $P(0 \le Z \le 1.62) \approx 0.4474$

3. $P(-2.1 < Z < -1.2) = P(1.2 < Z < 2.1) = P(0 \le Z < 2.1) - P(0 \le Z \le 1.2)$

$\approx 0.4821 - 0.3849 = 0.0972$

5. $P(-0.66 \le Z \le 1.32) = P(-0.66 \le Z < 0) + P(0 \le Z \le 1.32)$

$= P(0 < Z \le 0.66) + P(0 \le Z \le 1.32)$

$\approx 0.2454 + 0.4066 = 0.6520$

7. $P(Z > 3.04) = P(0 \le Z) - P(0 \le Z \le 3.04)$

$\approx 0.5 - 0.4988 = 0.0012$

9. $P(Z \leq 2.82) = P(Z < 0) + P(0 \leq Z \leq 2.82)$
$$= 0.5 + 0.4976 = 0.9976$$

11. $P(0 < X < 8) = P\left(\dfrac{0-6}{2} < Z < \dfrac{8-6}{2}\right) = P(-3 < Z < 1)$
$$= P(-3 < Z < 0) + P(0 \leq Z < 1)$$
$$= P(0 < Z < 3) + P(0 \leq Z < 1)$$
$$\approx 0.4987 + 0.3413 = 0.8400$$

13. $P(2 \leq X) = P\left(\dfrac{2-3.4}{1.2} \leq Z\right) \approx P(-1.17 \leq Z)$
$$= P(-1.17 \leq Z < 0) + P(0 \leq Z)$$
$$= P(0 < Z \leq 1.17) + P(0 \leq Z)$$
$$\approx 0.3790 + 0.5 = 0.8790$$

15. $P(-3 \leq X \leq 0) = P\left(\dfrac{-3-(-5)}{8} \leq Z \leq \dfrac{0-(-5)}{8}\right) \approx P(0.25 \leq Z \leq 0.63)$
$$= P(0 \leq Z \leq 0.63) - P(0 \leq Z < 0.25)$$
$$\approx 0.2357 - 0.0987 = 0.1370$$

17. $P(X < 500) = P\left(Z < \dfrac{500-541.2}{32.7}\right) \approx P(Z < -1.26)$
$$= P(Z > 1.26)$$
$$= P(0 \leq Z) - P(0 \leq Z \leq 1.26)$$
$$\approx 0.5 - 0.3962 = 0.1038$$

19. $P(X \geq 2) = P\left(Z \geq \dfrac{2-2.88}{0.55}\right) = P(Z \geq -1.6)$
$$= P(-1.6 \leq Z < 0) + P(0 \leq Z)$$
$$= P(0 < Z \leq 1.6) + P(0 \leq Z)$$
$$\approx 0.4452 + 0.5 = 0.9452$$

21. The value of z corresponding to $0.6 - 0.5 = 0.1$ is 0.25. Therefore, the 60th percentile is 0.25.

23. The value of z corresponding to $0.5 - 0.15 = 0.35$ is 1.04. Thus, the 15th percentile is -1.04.

25. The value of z corresponding to $0.94 - 0.5 = 0.44$ is $(1.55 + 1.56)/2 = 1.555$. Thus, the 94th percentile is 1.555.

27. The value of z corresponding to $0.55 - 0.5 = 0.05$ is 0.13. Therefore, the 55th percentile is given by $2 + 0.13 \cdot 5 = 2.65$.

29. The value of z corresponding to $0.5 - 0.06 = 0.44$ is $(1.55 + 1.56)/2 = 1.555$, so $z = -1.555$. Therefore, the 6th percentile is given by $-3 - 1.555 \cdot 7 = -13.885$.

31. The value of z corresponding to $0.5 - 0.24 = 0.26$ is 0.71, so $z = -0.71$. Therefore, the 24th percentile is given by $100.9 - 0.71 \cdot 12.7 = 91.883$.

33.
$$P(X \leq 40) = P\left(Z \leq \frac{40 - 38}{7}\right) \approx P(Z \leq 0.29)$$
$$= P(Z \leq 0) + P(0 < Z \leq 0.29)$$
$$= 0.5 + 0.1141 = 0.6141.$$

Rounding, we see that 40 corresponds to the 61st percentile.

35.
$$P(X > 800) = P\left(Z > \frac{800 - 505}{111}\right) \approx P(Z > 2.66)$$
$$= P(0 \leq Z) - P(0 \leq Z \leq 2.66)$$
$$\approx 0.5 - 0.4961 = 0.0039 = 0.39\%$$

37.
$$P(110 \leq X \leq 120) = P\left(\frac{110 - 100}{16} \leq Z \leq \frac{120 - 100}{16}\right)$$
$$\approx P(0.63 \leq Z \leq 1.25)$$
$$= P(0 \leq Z \leq 1.25) - P(0 \leq Z < 0.63)$$
$$\approx 0.3944 - 0.2357 = 0.1587 = 15.87\%$$

39.
$$P(X < 120) = P\left(Z < \frac{120 - 100}{16}\right) = P(Z < 1.25)$$
$$= P(Z \leq 0) + P(0 \leq Z < 1.25)$$
$$\approx 0.5 + 0.3944 = 0.8944 = 89.44\%$$

41. The value of z corresponding to $0.72 - 0.5 = 0.22$ is 0.58. Therefore, the score that falls at the 72nd percentile is $100 + 0.58 \cdot 16 = 109.28 \approx 109$.

43.
$$P(X \leq 130) = P\left(Z \leq \frac{130 - 100}{16}\right) \approx P(Z \leq 1.88)$$
$$= P(Z \leq 0) + P(0 < Z \leq 1.88)$$
$$\approx 0.5 + 0.4699 = 0.9699$$

Rounding, we see that a score of 130 corresponds to the 97th percentile.

45.
$$P(66 \leq X \leq 68) = P\left(\frac{66 - 70}{2} \leq Z \leq \frac{68 - 70}{2}\right)$$
$$= P(-2 \leq Z \leq -1) = P(1 \leq Z \leq 2)$$
$$= P(0 \leq Z \leq 2) - P(0 \leq Z < 1)$$
$$\approx 0.4772 - 0.3413 = 0.1359 = 13.59\%$$

47.
$$P(X > 76) = P\left(Z > \frac{76 - 70}{2}\right) = P(Z > 3)$$
$$= P(0 \leq Z) - P(0 \leq Z \leq 3)$$
$$\approx 0.5 - 0.4987 = 0.0013 = 0.13\%$$

49. The value of z corresponding to $0.5 - 0.1 = 0.4$ is 1.28, so $z = -1.28$. Therefore, the height of a man at the 10th percentile is $70 - 1.28 \cdot 2 = 67.44 \approx 67.4$ inches $= 5'\,7.4''$.

51.
$$P(X \le 60) = P\left(Z \le \frac{60 - 64.5}{2}\right) = P(Z \le -2.25)$$
$$= P(Z \ge 2.25)$$
$$= P(0 \le Z) - P(0 \le Z < 2.25)$$
$$\approx 0.5 - 0.4878 = 0.0122$$

Rounding, we see that a height of $5'$ corresponds to the 1st percentile.

53.
$$P(62 \le X \le 66) = P\left(\frac{62 - 64.5}{2} \le Z \le \frac{66 - 64.5}{2}\right)$$
$$= P(-1.25 \le Z \le 0.75)$$
$$= P(-1.25 \le Z < 0) + P(0 \le Z \le 0.75)$$
$$= P(0 < Z \le 1.25) + P(0 \le Z \le 0.75)$$
$$\approx 0.3944 + 0.2734 = 0.6678 = 66.78\%$$

55. The value of z corresponding to $0.5 - 0.05 = 0.45$ is $(1.64 + 1.65)/2 = 1.645$, so $z = -1.645$. Therefore, the height of a woman at the 5th percentile is $64.5 - 1.645 \cdot 2 = 61.21$ inches $\approx 5'\,1.2''$.

57. (a) Because only 10% of the scores are given a grade of A, the minimum score required to earn an A is the score that lies at the 90th percentile. The value of z corresponding to $0.9 - 0.5 = 0.4$ is 1.28, so the score at the 90th percentile is $68 + 1.28 \cdot 14 = 85.92 \approx 86$. Therefore, 86 is the minimum score required to earn an A.

(b) The minimum score required to earn a B is the score that lies at the 65th percentile. The value of z corresponding to $0.65 - 0.5 = 0.15$ is 0.39, so the score at the 65th percentile is $68 + 0.39 \cdot 14 = 73.46 \approx 73$. Therefore, 73 is the minimum score required to earn a B.

(c) The minimum score required to pass is the score that lies at the 10th percentile. The value of z corresponding to $0.5 - 0.1 = 0.4$ is 1.28, so $z = -1.28$. Thus, the score at the 10th percentile is $68 - 1.28 \cdot 14 = 50.08 \approx 50$. Therefore, 50 is the minimum score required to pass.

59.
$$P(18 \le X \le 22) = P\left(\frac{18 - 27.5}{6} \le Z \le \frac{22 - 27.5}{6}\right)$$
$$\approx P(-1.58 \le Z \le -0.92)$$
$$= P(0.92 \le Z \le 1.58)$$
$$= P(0 \le Z \le 1.58) - P(0 \le Z < 0.92)$$
$$\approx 0.4429 - 0.3212 = 0.1217 \approx 12.17\%$$

61. The value of z corresponding to $0.6 - 0.5 = 0.1$ is 0.25. Therefore, the age that falls at the 60th percentile is $27.5 + 0.25 \cdot 6 = 29$ years old.

63.
$$P(X \ge .347) = P\left(Z \ge \frac{.347 - .269}{.031}\right) \approx P(Z \ge 2.52)$$
$$= P(0 \le Z) - P(0 \le Z < 2.52)$$
$$\approx 0.5 - 0.4941 = 0.0059 \approx 0.59\%$$

65. $P(X \le .197) = P\left(Z \le \dfrac{.197 - .269}{.031}\right) \approx P(Z \le -2.32)$

$\qquad = P(Z \ge 2.32)$

$\qquad = P(0 \le Z) - P(0 \le Z < 2.32)$

$\qquad \approx 0.5 - 0.4898 = 0.0102 = 1.02\%$

67. $P(X > 25) = P\left(Z > \dfrac{25 - 32.4}{9.9}\right) \approx P(Z > -0.75)$

$\qquad = P(-0.75 < Z < 0) + P(0 \le Z)$

$\qquad = P(0 < Z < -0.75) + 0.5$

$\qquad \approx 0.2734 + 0.5 = 0.7734 = 77.34\%$

69. $P(X \le 30) = P\left(Z \le \dfrac{30 - 32.4}{9.9}\right) \approx P(Z \le -0.24)$

$\qquad = P(Z \ge 0.24)$

$\qquad = P(0 \le Z) - P(0 \le Z < 0.24)$

$\qquad \approx 0.5 - 0.0948 = 0.4052$

Rounding, we see that a 30-year-old inmate falls at the 41st percentile.

71. $P(5000 \le X \le 6000) = P\left(\dfrac{5000 - 4016}{654} \le Z \le \dfrac{6000 - 4016}{654}\right)$

$\qquad \approx P(1.50 \le Z \le 3.03)$

$\qquad = P(0 \le Z \le 3.03) - P(0 \le Z < 1.50)$

$\qquad \approx 0.4988 - 0.4332 = 0.0656 = 6.56\%$

73. $P(X \le 3000) = P\left(Z \le \dfrac{3000 - 4016}{654}\right) \approx P(Z \le -1.55)$

$\qquad = P(Z \ge 1.55)$

$\qquad = P(0 \le Z) - P(0 \le Z < 1.55)$

$\qquad \approx 0.5 - 0.4394 = 0.0606$

Rounding, we see that a 3000-pound truck lies at the 6th percentile.

75. $P(258 \le X \le 286) = P\left(\dfrac{258 - 272}{17} \le Z \le \dfrac{286 - 272}{17}\right)$

$\qquad = P\left(\dfrac{-14}{17} \le Z \le \dfrac{14}{17}\right)$

$\qquad \approx P(-0.82 \le Z \le 0.82)$

$\qquad = 2P(0 \le Z \le 0.82)$

$\qquad = 2 \cdot (0.2939) = 0.5878 = 58.78\%$

77. Because 7% of the scores are 90 or above, 90 falls at the 93rd percentile. The value of z corresponding to $0.93 - 0.5 = 0.43$ is 1.48, and therefore a score of 90 corresponds to a z-score of 1.48. Because 12% of the scores are 50 or below, 50 falls at the 12th percentile. The value of z corresponding

$0.5 - 0.12 = 0.38$ is 1.175, so a score of 50 corresponds to a z-score of -1.175. It follows that $90 - 50 = 40$ is $1.48 - (-1.175) = 2.655$ standard deviations. Therefore, the standard deviation is given by

$$\sigma = \frac{40}{2.655} \approx 15.1.$$

Finally, because a score of 90 is 1.48 standard deviations above the mean, we see that the mean is given by

$$\mu = 90 - 1.48 \cdot 15.1 = 67.652 \approx 67.7.$$

SECTION 5.5

1. For $n = 50$ and confidence level 60%, the margin of error is $0.84 \cdot 10/\sqrt{50} \approx 1.19$. Similarly, we compute the other table entries and get:

Confidence Level

n	60%	70%	80%	90%
50	1.19	1.47	1.81	2.33
100	0.84	1.04	1.28	1.65
500	0.38	0.47	0.57	0.74
1000	0.27	0.33	0.40	0.52

[handwritten: $0.84 \cdot 10 = 8.4$]

[handwritten: $8.4 \div \sqrt{50} = 1.19$]

3. (a) margin of error $= 1.88 \cdot 0.078/\sqrt{50} \approx 0.021$ mg/l

   (b) The confidence interval is from $0.20 - 0.021 = 0.179$ to $0.20 + 0.021 = 0.221$ mg/l.

5. The margin of error is $2.05 \cdot 26.6/\sqrt{288} \approx 3.2$ ppm. The confidence interval is from $660.5 - 8.2 = 657.3$ to $660.5 + 8.2 = 663.7$ ppm.

7. The margin of error is $1.44 \cdot 0.47/\sqrt{323} \approx 0.04$ g/cc. The confidence interval is from $2.74 - 0.04 = 2.70$ to $2.74 + 0.04 = 2.78$ g/cc.

9. The margin of error is $2.17 \cdot 0.23/\sqrt{107} \approx 0.05$ g/cc. The confidence interval is from $2.22 - 0.05 = 2.17$ to $2.22 + 0.05 = 2.27$ g/cc.

11. The margin of error is $1.28 \cdot 2.1/\sqrt{113} \approx 0.3$ months. The confidence interval is from $1.9 - 0.3 = 1.6$ to $1.9 + 0.3 = 2.2$ months.

13. The margin of error is $1.75 \cdot 4.43/\sqrt{68} \approx 0.94$ μg/g. The confidence interval is from $4.05 - 0.94 = 3.11$ to $4.05 + 0.94 = 4.99$ μg/g.

15. (a) The margin of error is $2.575 \cdot 0.9/\sqrt{324} \approx 0.1$ g/dl. The confidence interval is from $11.8 - 0.1 = 11.7$ to $11.8 + 0.1 = 11.9$ g/dl.

   (b) The margin of error is $2.575 \cdot 1.7/\sqrt{258} \approx 0.3$ g/dl. The confidence interval is from $11.1 - 0.3 = 10.8$ to $11.1 + 0.3 = 11.4$ g/dl.

   (c) Yes, because the confidence intervals in parts (a) and (b) do not overlap at all.

   (d) $P(X < 9.8) = P\left(Z < \dfrac{9.8 - 11.8}{0.9}\right) \approx P(Z < -2.22)$

   $= P(Z > 2.22) = P(0 \le Z) - P(0 \le Z \le 2.22)$

   $\approx 0.5 - 0.4868 = 0.0132 = 1.32\%$

(e) $P(X < 9.8) = P\left(Z < \dfrac{9.8 - 11.1}{1.7}\right) \approx P(Z < -0.76)$

$= P(Z > 0.76) = P(0 \le Z) - P(0 \le Z \le 0.76)$

$\approx 0.5 - 0.2764 = 0.2236 = 22.36\%$

(f) The answers to parts (d) and (e) seem close enough to the actual percentages to support an assumption that the distributions are normal.

17. The sample mean is $2749/30 \approx 91.63$, and the sample standard deviation is $\sqrt{10{,}678.966667/29} \approx 19.19$. The margin of error is $1.96 \cdot 19.19/\sqrt{30} \approx 6.87$. The confidence interval is from $91.63 - 6.87 = 84.76\%$ to $91.63 + 6.87 = 98.50\%$.

19. We agree with the conclusion because both confidence intervals are under 100 percent and sodium content is never more than 30% above what is labeled.

21. Setting the margin of error formula equal to 1 and solving for n we get

$$\frac{1.96 \cdot 19.19}{\sqrt{n}} = 1$$

$$1.96 \cdot 19.19 = \sqrt{n}$$

$$\sqrt{n} = 37.6124$$

$$n = (37.6124)^2 \approx 1414.69.$$

Therefore, a sample size of 1415 or more is needed.

23. The sample mean is

$$\bar{x} = \frac{7 \cdot 0 + 33 \cdot 1 + 30 \cdot 2 + 10 \cdot 3 + 5 \cdot 4 + 2 \cdot 5 + 1 \cdot 6 + 1 \cdot 8}{7 + 33 + 30 + 10 + 5 + 2 + 1 + 1}$$

$$= \frac{167}{89} = 1.88 \text{ siblings.}$$

The sample standard deviation is

$$s = \sqrt{\frac{7 \cdot (0 - \bar{x})^2 + 33 \cdot (1 - \bar{x})^2 + 30 \cdot (20 - \bar{x})^2 + 10 \cdot (3 - \bar{x})^2 + 5 \cdot (4 - \bar{x})^2 + 2 \cdot (5 - \bar{x})^2 + 1 \cdot (6 - \bar{x})^2 + 1 \cdot (8 - \bar{x})^2}{88}}$$

$$\approx 1.35 \text{ siblings.}$$

The margin of error is $1.75 \cdot 1.35/\sqrt{89} \approx 0.25$ siblings. The confidence interval is from $1.88 - 0.25 = 1.63$ to $1.88 + 0.25 = 2.13$ siblings.

25. The sample standard deviation is $s = 224 \cdot \sqrt{151/150} \approx 225\,\mu\text{g/l}$. The margin of error is $2.33 \cdot 225/\sqrt{151} \approx 43\,\mu\text{g/l}$. The confidence interval is from $342 - 43 = 299$ to $342 + 43 = 385\,\mu\text{g/l}$.

27. The sample standard deviation is $s = 19 \cdot \sqrt{255/254} \approx 19$ degrees. The margin of error is $2.33 \cdot 19/\sqrt{255} \approx 3$ degrees. The confidence interval is from $124 - 3 = 121$ to $124 + 3 = 127$ degrees.

SECTION 5.6

1. For $n = 50$ and confidence level 60%, the margin of error is $0.84(0.5)/\sqrt{50} \approx 0.0594 = 5.94\%$. Similarly, we compute the other table entries and get:

Confidence Level

n	60%	70%	80%	90%
50	5.94%	7.35%	9.05%	11.63%
100	4.20%	5.20%	6.40%	8.23%
500	1.88%	2.33%	2.86%	3.68%
1000	1.33%	1.64%	2.02%	2.60%

3. margin of error $= 1.96(0.5)/\sqrt{1126} \approx 0.03 = 3\%$

5. (a) margin of error $= 2.05(0.5)/\sqrt{1530} \approx 0.0262 = 2.62\%$

    (b) One possible explanation is that pennies are less recognizable with tails up and therefore more likely to be left behind.

7. We first note that $259/1003 \approx 0.258 = 25.8\%$ of the households used the Internet or World Wide Web. The margin of error is $217(0.5)/\sqrt{1003} \approx 0.034 = 3.4\%$. The confidence interval is from $25.8\% - 3.4\% = 22.4\%$ to $25.8\% + 3.4\% = 29.2\%$.

9. The margin of error is $1.70(0.5)/\sqrt{120} \approx 0.078 = 7.8\%$. The confidence interval is from $62.5\% - 7.8\% = 54.7\%$ to $62.5\% + 7.8\% = 70.3\%$.

11. Setting the margin of error equal to 0.04 and solving for n we get

$$\frac{2.575(0.5)}{\sqrt{n}} = 0.04$$

$$2.575(0.5) = 0.04\sqrt{n}$$

$$\sqrt{n} = \frac{2.575(0.5)}{0.04}$$

$$\sqrt{n} = 32.1875$$

$$n = (32.1875)^2 \approx 1036.04.$$

Therefore, at least 1037 people should be included in the survey.

13. Setting the margin of error equal to 0.035 and solving for n we get

$$\frac{1.96(0.5)}{\sqrt{n}} = 0.035$$

$$1.96(0.5) = 0.035\sqrt{n}$$

$$\sqrt{n} = \frac{1.96(0.5)}{0.035}$$

$$\sqrt{n} = 28$$

$$n = (28)^2 = 784.$$

Therefore, there were approximately 784 respondents.

15. Setting the margin of error equal to 0.044 and solving for z we find

$$\frac{z(0.5)}{\sqrt{500}} = 0.044$$

$$z = \frac{0.044\sqrt{500}}{0.5} \approx 1.97.$$

Therefore, the confidence level is approximately $2(0.4756) = 0.9512 \approx 0.95 = 95\%$.

17. Setting the margin of error equal to 0.03 and solving for z we find

$$\frac{z(0.5)}{\sqrt{1004}} = 0.03$$

$$z = \frac{0.03\sqrt{1004}}{(0.5)} \approx 1.90.$$

Therefore, the confidence level is approximately $2(0.4713) = 0.9426 \approx 0.94 = 94\%$. Because 95% is a typical confidence level, it might be the one actually used in the poll. To check whether 95% is reasonable, we compute the margin of error for the 95% confidence level and get $1.96(0.5)/\sqrt{1004} \approx 0.03092$, which would round to the given 3% margin of error.

19. First, we solve for z in the margin of error formula for the poll of all 800 respondents and find

$$\frac{z(0.5)}{\sqrt{800}} = 0.035$$

$$z = \frac{0.035\sqrt{800}}{0.5} \approx 1.979899.$$

Now, letting $z = 1.979899$ in the margin of error formula for the poll of only the residents of Pima County and solving for n we get

$$\frac{(1.979899)(0.5)}{\sqrt{n}} = 0.08$$

$$(1.979899)(0.5) = 0.08\sqrt{n}$$

$$\sqrt{n} = \frac{(1.979899)(0.5)}{0.08}$$

$$\sqrt{n} \approx 12.374369$$

$$n \approx (12.374369)^2 \approx 153.$$

Therefore, there were approximately 153 residents of Pima County included in the poll.

21. Because any change in one candidate's proportion in a two-candidate race changes the difference in proportion by twice the change, the margin of error for the difference is double the margin of error for either candidate.

SECTION 5.7

1. The overall unemployment rate cannot be higher than both the unemployment rate for men and the unemployment rate for women.

3. It used to be that many children were actively discouraged from being left-handers. Children from this era are older members of today's population, so should have fewer left-handers. Attitudes have changed, so that almost all children today are encouraged to favor their most natural hand. Those who were children recently are still quite young.

5. Birthday celebrations may involve more activity than that to which people are accustomed and would therefore raise the rate of heart attacks.

7. The order should matter more for long and involved choices. A subject's memory of earlier choices might fade in a telephone interview, whereas the subject's attention might drift in the later choices of a written questionnaire.

Chapter 5 Review Exercises

1. (a) There are 15 tournaments in all.

Country	Frequency	Relative Frequency
Italy	3	0.2000
Uruguay	1	0.0667
Germany	3	0.2000
Brazil	4	0.2667
England	1	0.0667
Argentina	2	0.1333
France	1	0.0667

(b)

(c)

Country	Angle Measure (degrees)
Italy	72
Uruguay	24
Germany	72
Brazil	96
England	24
Argentina	48
France	24

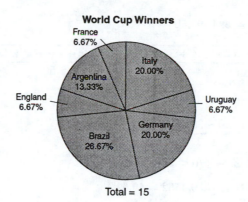

World Cup Winners

Total = 15

3. There were 30 NFL teams in the 1997-1998 season.

Ticket Price	Frequency	Relative Frequency
29.5 – 39.5	16	0.5333
39.5 – 49.5	7	0.2333
49.5 – 59.5	5	0.1667
59.5 – 69.5	1	0.0333
69.5 – 79.5	1	0.0333

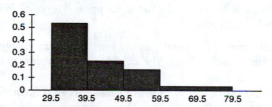

1997–1998 NFL Average Ticket Price (dollars)

5. **(a)** $\mu = \frac{1\cdot21.05+3\cdot21.85+1\cdot22.05+1\cdot22.45+1\cdot22.65+2\cdot23.05+1\cdot23.25+1\cdot23.45+3\cdot24.05+1\cdot24.85}{15}$

$$= \frac{343.55}{15} \approx 22.90 \, \text{mm}$$

(b) $\mu = \frac{1\cdot19.85+1\cdot20.05+1\cdot20.25+3\cdot20.85+3\cdot21.05+1\cdot21.25+1\cdot21.45+3\cdot22.05+1\cdot22.25}{15}$

$$= \frac{316.95}{15} = 21.13 \, \text{mm}$$

(c) range $= 24.85 - 21.05 = 3.8$ mm

$$\sigma = \sqrt{\frac{1\cdot(21.05-\mu)^2 + 3\cdot(21.05-\mu)^2 + \cdots + 1\cdot(24.85-\mu)^2}{15}} \approx 1.03 \, \text{mm}$$

(d) range $= 22.25 - 19.85 = 2.4$ mm

$$\sigma = \sqrt{\frac{1\cdot(19.85-\mu)^2 + 1\cdot(20.05-\mu)^2 + \cdots + 1\cdot(22.25-\mu)^2}{15}} \approx 0.72 \, \text{mm}$$

7. (a) $\mu \approx \dfrac{1 \cdot 54{,}999.5 + 3 \cdot 64{,}999.5 + 8 \cdot 74{,}999.5 + 10 \cdot 84{,}999.5 + 12 \cdot 94{,}999.5 + 5 \cdot 104{,}999.5 + 5 \cdot 114{,}999.5 + 4 \cdot 124{,}999.5 + 2 \cdot 134{,}999.5}{1 + 3 + 8 + 10 + 12 + 5 + 5 + 4 + 2}$

$$= \frac{4{,}709{,}975}{50} = \$94{,}199.50$$

  (b) $\sigma \approx \sqrt{\dfrac{1 \cdot (54{,}999.5 - \mu)^2 + 3 \cdot (64{,}999.5 - \mu)^2 + \cdots + 4 \cdot (124{,}999.5 - \mu)^2 + 2 \cdot (134{,}999.5 - \mu)^2}{50}}$

$$\approx \$18{,}851$$

9. $P(X > 40) = P\left(Z > \dfrac{40 - 38.6}{13.1}\right) \approx P(Z > 0.11)$

$$= P(0 \leq Z) - P(0 \leq Z \leq 0.11)$$

$$\approx 0.5 - 0.0438 = 0.4562 = 45.62\%$$

11. $P(X \leq 30) = P\left(Z \leq \dfrac{30 - 38.6}{13.1}\right) \approx P(Z \leq -0.66)$

$$= P(Z \geq 0.66) = P(0 \leq Z) - P(0 \leq Z < 0.66)$$

$$\approx 0.5 - 0.2454 = 0.2546$$

Rounding, we see that age 30 falls at the 25th percentile.

13. (a) The margin of error is $1.70 \cdot 4.02/\sqrt{162} \approx 0.54$ ppm. The confidence interval is from $3.06 - 0.54 = 2.52$ ppm to $3.06 + 0.54 = 3.60$ ppm.

  (b) The margin of error is $1.70 \cdot 50.2/\sqrt{162} \approx 6.7$ ppm. The confidence interval is from $36.0 - 6.7 = 29.3$ to $36.0 + 6.7 = 42.7$ ppm.

  (c) Selenium content seems to depend heavily on the region, as evidenced by the gap between the confidence interval in part (a) and the confidence interval in part (b).

15. We first note that $769/1000 = 0.769 = 76.9\%$ of those surveyed favored the death penalty. The margin of error is $2.33(0.5)/\sqrt{1000} \approx 0.037 = 3.7\%$. The confidence interval is from $76.9\% - 3.7\% = 73.2\%$ to $76.9\% + 3.7\% = 80.6\%$.

17. Setting the margin of error equal to 50,000 and solving for n we get

$$\frac{1.96 \cdot 476{,}984}{\sqrt{n}} = 50{,}000$$

$$1.96 \cdot 476{,}984 = 50{,}000\sqrt{n}$$

$$\sqrt{n} = \frac{1.96 \cdot 476{,}984}{50{,}000}$$

$$\sqrt{n} \approx 18.697773$$

$$n \approx (18.697773)^2 \approx 349.61.$$

Therefore, at least 350 cases would be needed.

CHAPTER 6
Student Solution Manual

SECTION 6.1

1. A, B

3. B, C, D, E

5. It has an Eulerian circuit. One possibility is as follows:

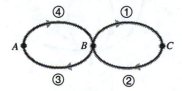

7. It has four odd vertices, A, B, C, and D, so it has neither an Eulerian circuit nor an Eulerian path.

9. It has an Eulerian circuit. One possibility is as follows:

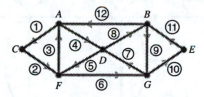

11. It has two odd vertices, A and D, so it does not have an Eulerian circuit. It does have an Eulerian path. One possibility is as follows:

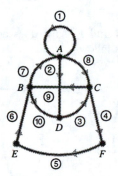

13. It has two odd vertices, A and C, so it does not have an Eulerian circuit. It does have an Eulerian path. One possibility is as follows:

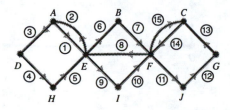

15. The corresponding graph has two odd vertices, B and D, so it is possible to take such a walk. One such walk is as follows:

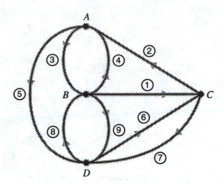

17. All the vertices of the corresponding graph are even, so the graph has an Eulerian circuit. One possibility is as follows:

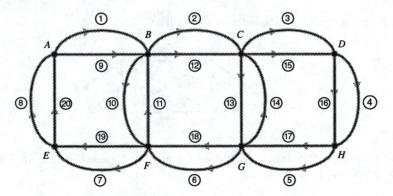

19. The graph has two odd vertices, B and F, so it does not have an Eulerian circuit. It does have an Eulerian path. One possibility is as follows:

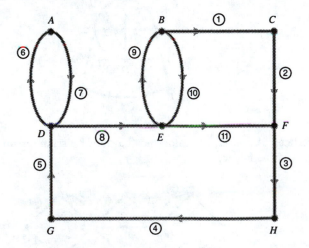

21. (a)

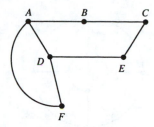

(b) The graph has two odd vertices, A and D, so it is not possible.

23. One possible eulerization is as follows:

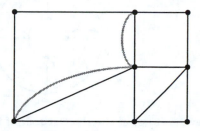

25. One possible eulerization is as follows:

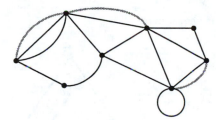

27. One possible eulerization is as follows:

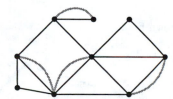

29. One possible eulerization is as follows:

31. (a)

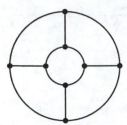

(b) One possible eulerization is as follows:

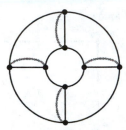

(c) The eulerization shown in (b) gives the shortest tour of the park.

SECTION 6.2

1. One Hamiltonian circuit is shown here.

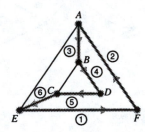

3. One Hamiltonian circuit is shown here.

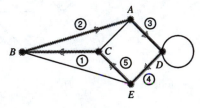

5. There are no Hamiltonian circuits on this graph because any circuit would have to go through the edge between vertices D and E twice and therefore visit both vertex D and vertex E twice.

7. **(a)**

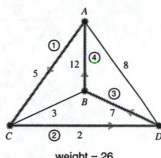

weight = 26

(b) The edges are chosen in the order: CD, BC, AD, AB.

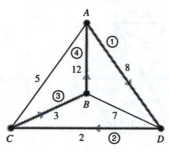

or the reverse circuit

weight = 25

9. **(a)**

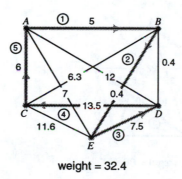

weight = 32.4

or

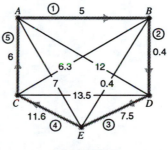

weight = 30.5

(b) The edges are chosen in the order: BD, BE, AC, AE, CD or BE, BD, AC, AE, CD.

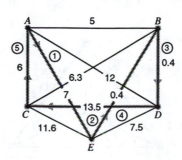

or the reverse circuit

weight = 27.3

11. (a)

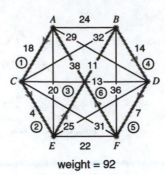

weight = 92

(b) The edges are chosen in the order: CE, DF, BE, CD, AB, AF.

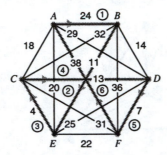

or the reverse circuit

weight = 97

13. (a)

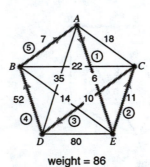

weight = 86

(b) The edges are chosen in the order: AE, AB, CD, CE, BD, and we get the same circuit as in part (a) or the reverse circuit; weight = 86.

15. (a)

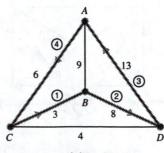

weight = 30

(b) The edges are chosen in the order: BC, CD, AB, AD.

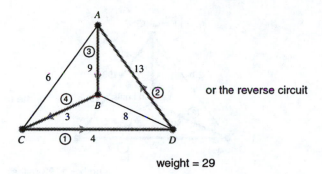

weight = 29

(c) After checking all possible circuits we find that the exact solution of the traveling salesman problem is as follows:

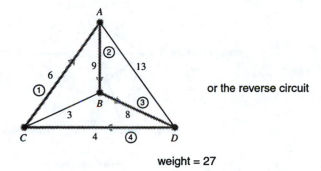

weight = 27

17. (a)

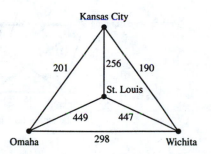

(b)

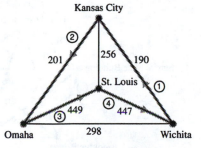

length = 1287 miles

(c) The edges are chosen in the order: K.C. to Wichita, K.C. to Omaha, St. Louis to Wichita, St. Louis to Omaha, and we get the same circuit as the one found in part (b) or the reverse circuit; length = 1287 miles.

(d)

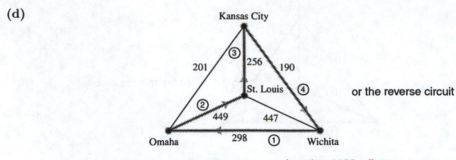

length = 1193 miles

This route is 94 miles shorter than those found in parts (b) and (c).

19. (a)

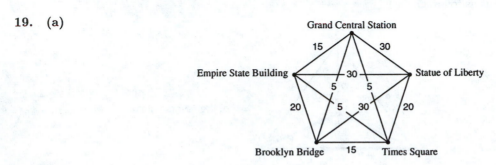

(b)

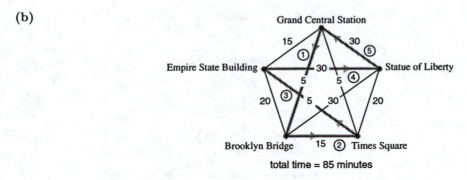

total time = 85 minutes

or

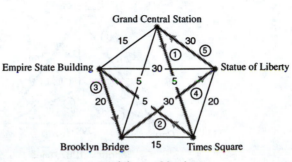

total time = 90 minutes

(c) The three edges of weight 5 are the first three chosen (in any order), the edges from the Empire State Building to the Statue of Liberty and from the Statue of Liberty to the Brooklyn Bridge are chosen next (in either order).

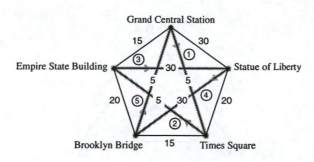

or the reverse circuit

total time = 75 minutes

21. **(a)**

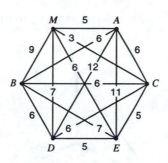

(b)

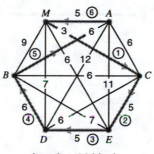

length = 30 blocks

(c) The edges are chosen in the order: first MC, then MA, CE, and DE in any order, and last AB and BD in either order. We get the same circuit as in part (b) or the reverse circuit; length = 30 blocks.

23. Base, Barnard 3, Shoemaker-Levy 7, Gunn, Swift, Neujmin 2, du Toit-Hartley, Harrington-Wilson, Base; total angular change = 419.9°.

25. **(a)**

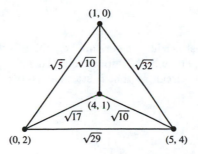

(b)

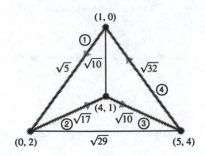

length $= \sqrt{5} + \sqrt{17} + \sqrt{10} + \sqrt{32} \approx 15.18$ cm

(c) The edges are chosen in the order: first $(1, 0)$ to $(0, 2)$, then $(1, 0)$ to $(4, 1)$ and $(4, 1)$ to $(5, 4)$ in either order, last $(0, 2)$ to $(5, 4)$.

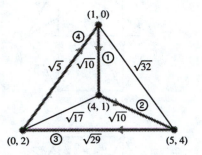

or the reverse circuit

length $= \sqrt{10} + \sqrt{10} + \sqrt{29} + \sqrt{5} \approx 13.95$ cm

27. One Hamiltonian circuit is as follows:

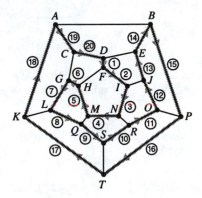

29. Application of the nearest neighbor algorithm beginning at vertices $A, B, C, D, E,$ and F yields circuits of weight 77, 75, 81, 75, 66, and 75, respectively. Therefore, the shortest circuit is the one starting at vertex E. Rewriting this circuit so that it starts at vertex B, we get the following circuit:

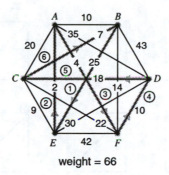

weight = 66

31. One possible answer is shown here.

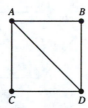

33. One possible answer is the following graph. The greedy algorithm gives the circuit shown, and this circuit has weight 16. (We have drawn the circuit starting at vertex A, but it has the same weight no matter at what vertex it starts.) All other Hamiltonian circuits on the graph, except the same circuit in reverse order, have weight 15.

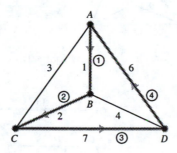

SECTION 6.3

1. not a tree because it includes a circuit

3. a tree, but not a spanning tree

5. a spanning tree

7. Three possibilities are as follows:

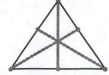 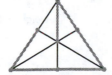 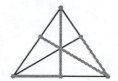

9.

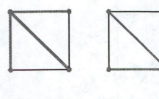

11.

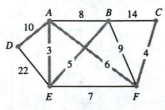

weight = 28

13.

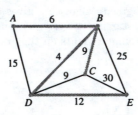 or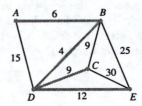

weight = 31

15.

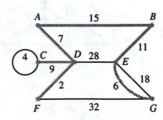

weight = 50

17.

weight = 47

19.

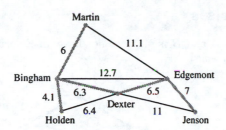

length = 29.9 miles

21. The flights below should be retained.

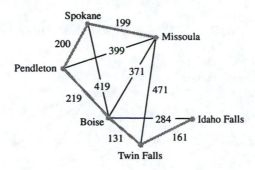

23. The cable should be installed as shown here.

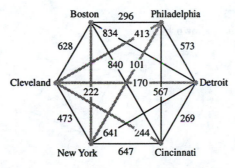

length = 1150 miles

25. The minimum cost network is shown here.

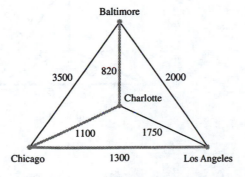

total monthly cost = $3220

Chapter 6 Review Exercises

1. It has two odd vertices, A and B, so it does not have an Eulerian circuit. It has an Eulerian path. One possibility is as follows:

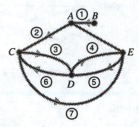

3. It has six odd vertices, A, B, C, E, F, and H, so it has neither an Eulerian circuit nor an Eulerian path.

5. One possible eulerization is shown here.

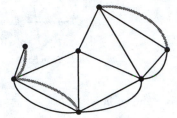

7. One Hamiltonian circuit is shown here.

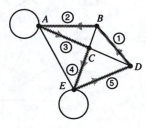

9. (a)

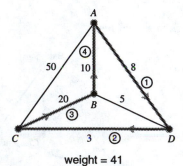

weight = 41

(b) The edges are chosen in the order: CD, BD, AB, AC.

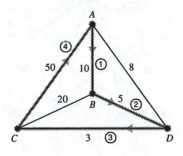

or the reverse circuit

weight = 68

11.

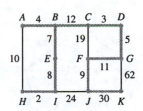

weight = 91

13. The graph has an Eulerian circuit. One possibility is as follows:

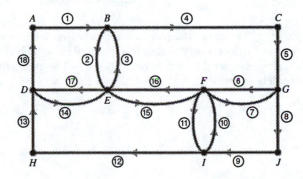

15. (a)

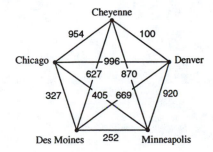

(b)

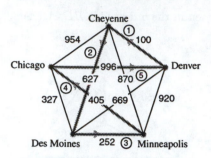

length = 2380 miles

(c) The edges are chosen in the order: Cheyenne to Denver, Des Moines to Minneapolis, Chicago to Des Moines, Cheyenne to Minneapolis, Chicago to Denver.

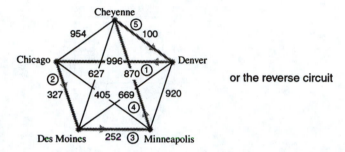

or the reverse circuit

length = 2545 miles

17. In a complete graph, every vertex is attached to every other vertex by exactly one edge. Therefore, in a complete graph with n vertices, every vertex has $n - 1$ edges attached to it. If n is even, then $n - 1$ is odd and hence every vertex is odd. If n is odd, then $n - 1$ is even and hence every vertex is even. It follows that a complete graph with n vertices has an Eulerian circuit if and only if n is odd.

CHAPTER 7

Student Solution Manual

SECTION 7.1

1. It is a polygon.

3. It is not a polygon because the line segments intersect at points other than the endpoints.

5. It is not a polygon because one of the sides is curved and therefore it is not made up of only line segments.

7. It is not a polygon because two of the endpoints of the line segments do not intersect other endpoints

9. concave; not regular; heptagon

11. convex; regular; nonagon

13. convex; not regular; quadrilateral

15. convex; regular; 11-gon

17. $(10-2)180° = 8 \cdot 180° = 1440°$

19. $(17-2)180° = 15 \cdot 180° = 2700°$

21. $(9-2)180° = 7 \cdot 180° = 1260°$

23. $180° - \dfrac{360°}{8} = 180° - 45° = 135°$

25. $180° - \dfrac{360°}{7} \approx 180° - 51.4286° = 128.5714°$

27. $180° - \dfrac{360°}{30} = 180° - 12° = 168°$

29. Let x be the measure of the unknown angle. Setting the sum of the angles of the given quadrilateral equal to the formula for the sum of the angles of any quadrilateral and solving for x, we find

$$54 + 44 + 241 + x = (4-2)180$$
$$339 + x = 360$$
$$x = 360 - 339 = 21°.$$

31. Let x be the measure of the unknown angles. Setting the sum of the angles of the given pentagon equal to the formula for the sum of the angles of any pentagon and solving for x, we find

$$35 + 42 + 93 + x + x = (5-2)180$$
$$170 + 2x = 540$$
$$2x = 540 - 170$$
$$2x = 370$$
$$x = \frac{370}{2} = 185°.$$

SECTION 7.2

1. It is not regular because it is not edge-to-edge.

3. It is not semiregular because not all vertices are of the same type.

5. It is not semiregular because the tiles are not all regular polygons and not all vertices are of the same type.

7. The four different vertex types are as follows:

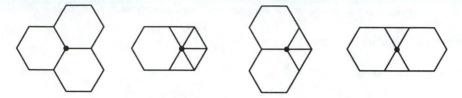

9. The two different vertex types are as follows:

11.

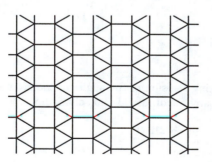

13. The only combination of regular polygons with five tiles about each vertex that does not include squares is one regular hexagon and four equilateral triangles. The resulting semiregular tiling is as follows:

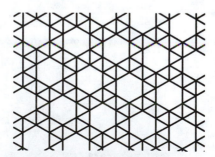

15. We continue the pattern by first placing a decagon at vertex A and then placing pentagons at vertices B and C. However, this results in three pentagons at vertex D (and a gap that cannot be filled), so the vertex type of D is not correct.

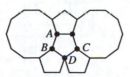

17. The measure of the angle of a regular 24-gon is $180° - 360°/24 = 180° - 15° = 165°$, so the sum of the angles of the remaining two polygons is $360° - 165° = 195°$. The only two regular polygons whose angle measures sum to $195°$ are a regular octagon with angle $180° - 360°/8 = 180° - 45° = 135°$ and an equilateral triangle with angle $60°$.

19.

21.

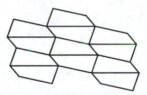

23.

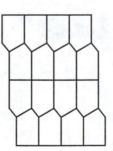

25. (a)

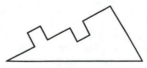

(b)

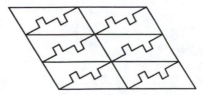

29.

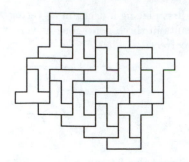

31.

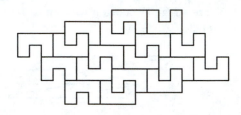

33. Two possible answers are as follows:

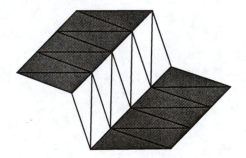

SECTION 7.3

1. convex

3. concave

5. $V - E + F = 10 - 17 + 9 = 2$

7. $V - E + F = 8 - 14 + 8 = 2$

9. $V - E + F = 4 - 6 + 4 = 2$

11. $V - E + F = 6 - 12 + 8 = 2$

13. $V - E + F = 12 - 30 + 20 = 2$

15. Substituting $E = 20$ and $V = 10$ into Euler's formula and solving for the number of faces, F, we get

$$10 - 20 + F = 2$$
$$-10 + F = 2$$
$$F = 2 + 10 = 12.$$

17. Substituting $F = 6$ and $V = 5$ into Euler's formula, we get

$$5 - E + 6 = 2$$
$$11 - E = 2$$
$$E = 11 - 2 = 9,$$

so the polyhedron has 9 edges. Because each of the 9 edges belongs to exactly 2 faces, the polygonal faces have a total of $9 \cdot 2 = 18$ sides altogether. Because each of the 6 polygonal faces must have at least 3 sides, we see that the only way the total number of sides can add up to 18 is to have 6 triangular faces. A sketch of one such convex polyhedron is shown here.

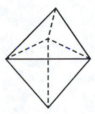

19. Substituting $E = 10$ and $V = 6$ into Euler's formula, we get

$$F - 10 + 6 = 2$$
$$F - 4 = 2$$
$$F = 2 + 4 = 6,$$

so the polyhedron has 6 faces. Because each of the 10 edges belongs to exactly 2 faces, the polygonal faces have a total of $10 \cdot 2 = 20$ sides altogether. Because each of the 6 polygonal faces must have at least 3 sides, we see that the only way the total number of sides can add up to 20 is to have 4 triangular faces and 2 quadrilateral faces or to have 5 triangular faces and 1 pentagonal face. A sketch of a convex polyhedron of each of these types is shown here.

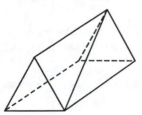 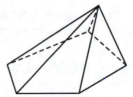

21. **(a)** The faces of the polyhedron have $4 \cdot 3 + 4 \cdot 4 = 28$ sides altogether. Because each edge belongs to two faces, the polyhedron has $28/2 = 14$ edges.

(b) Substituting $E = 14$ and $F = 8$ into Euler's formula and solving for the number of vertices, V, we get

$$V - 14 + 8 = 2$$
$$V - 6 = 2$$
$$V = 2 + 6 = 8.$$

(c)

23. $V - E + F = 16 - 24 + 10 = 2$

25. $V = 16$, $E = 32$, $F = 16$

27. $V = 40$, $E = 84$, $F = 40$

29. For the polyhedron with one hole in Exercise 25, we have $V - E + F = 16 - 32 + 16 = 0$. For the polyhedron with two holes in Exercise 26, we have $V - E + F = 28 - 58 + 28 = -2$. For the polyhedron with three holes in Exercise 27, we have $V - E + F = 40 - 84 + 40 = -4$. For the polyhedron with four holes in Exercise 27, we have $V - E + F = 52 - 110 + 52 = -6$. It appears that the sum $V - E + F$ is reduced by 2 for every hole. Therefore, the formula appears to be $V - E + F = 2 - 2h$, where h is the number of holes. (In fact, this conjectured formula is correct.)

31. **(a)** 8 hexagonal faces

(b) 6 square faces

33. 4 equilateral triangle faces and 4 regular hexagon faces

35. cube

37. dodecahedron

39. **(a)** Each of the p pentagonal faces has 5 vertices while each of the h hexagonal faces has 6 vertices, giving a total of $5p + 6h$ vertices. However, each vertex belongs to three faces, so the fullerene has $\frac{1}{3}(5p + 6h)$ vertices.

(b) Each of the p pentagonal faces has 5 sides while each of the h hexagonal faces has 6 sides, giving a total of $5p + 6h$ sides. However, each side belongs to two faces, so the fullerene has $\frac{1}{2}(5p + 6h)$ edges.

(c) Substituting $V = \frac{1}{3}(5p + 6h)$, $E = \frac{1}{2}(5p + 6h)$, and $F = p + h$ into Euler's formula and solving for p, we get

$$\tfrac{1}{3}(5p + 6h) - \tfrac{1}{2}(5p + 6h) + (p + h) = 2$$
$$\tfrac{5}{3}p + 2h - \tfrac{5}{2}p - 3h + p + h = 2$$
$$\left(\tfrac{5}{3} - \tfrac{5}{2} + 1\right)p = 2$$
$$\left(\tfrac{10}{6} - \tfrac{15}{6} + \tfrac{6}{6}\right)p = 2$$
$$\tfrac{1}{6}p = 2$$
$$p = 2 \cdot 6 = 12.$$

Therefore, any fullerene has exactly 12 pentagonal faces.

41. **(a)** Every edge belongs to exactly two faces of the polyhedron, and therefore the faces have a total of $2E$ sides. Because every face must have at least 3 sides, we see that

$$2E \geq 3F$$

$$E \geq \tfrac{3}{2}F.$$

(b) Every vertex of the polyhedron must have at least 3 edges meeting it and each edge belongs to exactly 2 vertices. It follows that

$$2E \geq 3V$$

$$E \geq \tfrac{3}{2}V.$$

43. Using the equation $E \geq \tfrac{3}{2}F$ and $F = 8$, we can conclude that $E \geq \tfrac{3}{2} \cdot 8 = 12$. Using Euler's formula with $F = 8$ and the equation $E \geq \tfrac{3}{2}V$, we find

$$V - E + 8 = 2$$

$$8 - 2 = E - V$$

$$6 = E - V$$

$$6 \geq \tfrac{3}{2}V - V$$

$$6 \geq \tfrac{1}{2}V$$

$$12 \geq V.$$

From the equation $E - V = 6$, we see that $E = V + 6$. Combining this fact with the conditions $V \leq 12$ and $E \geq 12$, it follows that the only possible values for V are $V = 6, 7, 8, 9, 10, 11,$ or 12. Using the equation $E = V + 6$, we find that the only pairs of values of E and V for which there may be a convex polyhedron with eight faces are:

$$V = 6, \ E = 12; \ V = 7, \ E = 13; \ V = 8, \ E = 14; \ V = 9, \ E = 15; \ V = 10, \ E = 16;$$

$$V = 11, \ E = 17; \ V = 12, \ E = 18.$$

Chapter 7 Review Exercises

1.

3. $180° - \dfrac{360°}{15} = 180° - 24° = 156°$

5.

7. The three different vertex types are shown here.

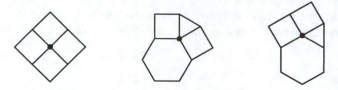

9. The measure of the angle of a regular 42-gon is

$$180° - 360°/42 = 180° - 8\tfrac{24}{42}° = 180° - 8\tfrac{4}{7}° = 171\tfrac{3}{7}°,$$

so the sum of the angles of the remaining two polygons is $360° - 171\tfrac{3}{7}° = 188\tfrac{4}{7}°$. The only two regular polygons whose angle measures sum to $188\tfrac{4}{7}°$ are a regular heptagon with angle $180° - 360°/7 = 180° - 51\tfrac{3}{7}° = 128\tfrac{4}{7}°$ and an equilateral triangle with angle $60°$.

11. One of the vertices is surrounded by a pentagon, an octagon, and a hexagon. If they were all regular, the sum of the angles would be

$$\left(180° - \tfrac{360°}{5}\right) + \left(180° - \tfrac{360°}{8}\right) + \left(180° - \tfrac{360°}{6}\right) = 108° + 135° + 120° = 363°.$$

Because the angles do not sum to $360°$, these polygons cannot all be regular.

13. Substituting $F = 14$ and $V = 13$ into Euler's formula and solving for the number of edges, E, we get

$$13 - E + 14 = 2$$
$$27 - E = 2$$
$$E = 27 - 2 = 25.$$

15. Because each of the 8 edges belongs to exactly 2 faces, the polygonal faces have a total of $8 \cdot 2 = 16$ sides altogether. Because each of the 5 polygonal faces must have at least 3 sides, we see that the only way the total number of sides can add up to 16 is to have 4 triangular faces and one quadrilateral face. A sketch of one such convex polyhedron is shown here.

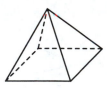

CHAPTER 8
Student Solution Manual

SECTION 8.1

1. True, because $42 = 7 \cdot 6$.

3. False, because $40/12 \approx 3.33$ is not an integer.

5. False, because $28/0$ is undefined.

7. False, because $-34/6 \approx -5.67$ is not an integer.

9. True, because $0 = 4 \cdot 0$.

11. True, because $240 = (-10)(-24)$.

13. $132 = 2^2 \cdot 3 \cdot 11$

15. $1625 = 5^3 \cdot 13$

17. $6800 = 2^4 \cdot 5^2 \cdot 17$

19. Carrying out the Sieve of Eratosthenes for all integers between 2 and 100, we are left with the following list of all primes less than or equal to 100: 2, 3, 5, 7, 11, 13, 17, 19, 23, 29, 31, 37, 41, 43, 47, 53, 59, 61, 67, 71, 73, 79, 83, 89, 97.

21. Because $\sqrt{751} \approx 27.40$, we need only check for divisibility by the primes 2, 3, 5, 7, 11, 13, 17, 19, and 23. We find that 751 is not divisible by any of these primes and is therefore prime.

23. Because $\sqrt{1817} \approx 42.63$, we need only check for divisibility by the primes 2, 3, 5, 7, 11, 13, 17, 19, 23, 29, 31, 37, and 41. We find that $1817 = 23 \cdot 79$, and therefore 1817 is not prime.

25. Because $\sqrt{2579} \approx 50.78$, we need only check for divisibility by the primes 2, 3, 5, 7, 11, 13, 17, 19, 23, 29, 31, 37, 41, 43, and 47. We find that 2579 is not divisible by any of these primes and is therefore prime.

27. Because $F_3 = 2^{2^3} + 1 = 257$ and $\sqrt{257} \approx 16.03$, we need only check for divisibility by the primes 2, 3, 5, 7, 11, and 13. We find that 257 is not divisible by any of these primes and is therefore prime.

29. $3 + 13$ or $5 + 11$

31. $3 + 43$ or $5 + 41$ or $17 + 29$ or $23 + 23$

33. $19 + 79$ or $31 + 67$ or $37 + 61$

35. One possibility is $86 = 7 + 13 + 23 + 43$.

37. One possibility is $232 = 3 + 11 + 19 + 31 + 37 + 41 + 43 + 47$.

39. $q = 9, r = 5$

41. $q = 17, r = 0$

43. $q = -8, r = 4$

45. $q = -14, r = 3$

47. $q = -9, r = 3$

49. From the prime factorizations $9 = 3^2$ and $15 = 3 \cdot 5$, we see that $\gcd(9, 15) = 3$.

51. $\gcd(0, 24) = 24$

53. From the prime factorizations $16 = 2^4$ and $40 = 2^3 \cdot 5$, we see that $\gcd(-16, 40) = 2^3 = 8$.

55. From the prime factorizations $77 = 7 \cdot 11$ and $130 = 2 \cdot 5 \cdot 13$, we see that $\gcd(77, 130) = 1$.

57. From the prime factorizations $650 = 2 \cdot 5^2 \cdot 13$ and $475 = 5^2 \cdot 19$, we see that $\gcd(650, 475) = 5^2 = 25$.

59. From the prime factorizations $525 = 3 \cdot 5^2 \cdot 7$ and $231 = 3 \cdot 7 \cdot 11$, we see that $\gcd(-525, -231) = 3 \cdot 7 = 21$.

61. The positive divisors of 10, other than 10 itself, are 1, 2, and 5. Because $1 + 2 + 5 = 8 \neq 10$, 10 is not perfect.

63. The positive divisors of 496, other than 496 itself, are 1, 2, 4, 8, 16, 31, 62, 124, and 248. Because

$$1 + 2 + 4 + 8 + 16 + 31 + 62 + 124 + 248 = 496,$$

we see that 496 is perfect.

65. (a) For every positive integer n, either n is even or $n+1$ is even. Therefore, $n(n+1)/2$ is an integer for every positive integer n.

(b) The number $n(n+1)/2$ is an even integer if and only if $n(n+1)$ is divisible by 4. Because exactly one of n and $n+1$ will be even, it follows that $n(n+1)/2$ is an even integer if and only if either n is divisible by 4 or $n+1$ is divisible by 4, in other words if either $n = 4k$ or $n = 4k - 1$, where k is positive integer.

SECTION 8.2

1. $48 = 5 \cdot 9 + 3$, so 48 mod 9 = 3.

3. $-38 \equiv -38 + 4 \cdot 12 \equiv 10 \pmod{12}$, so -38 mod 12 = 10.

5. $396 = 56 \cdot 7 + 4$, so 396 mod 7 = 4.

7. $14 = 7 \cdot 2$, so 14 mod 2 = 0.

9. $-342 = -57 \cdot 6$, so -342 mod 6 = 0.

11. $-54 \equiv -54 + 14 \cdot 4 \equiv 2 \pmod{4}$, so -54 mod 4 = 2.

13. $45{,}237 = 4523 \cdot 10 + 7$, so 45,237 mod 10 = 7.

15. $-252 \equiv -252 + 23 \cdot 11 \equiv 1 \pmod{11}$, so -252 mod 11 = 1.

17. $2295 = 143 \cdot 16 + 7$, so 2295 mod 16 = 7.

19. $-30{,}881 \equiv -30{,}881 + 618 \cdot 50 \equiv 19 \pmod{50}$, so $-30{,}881 \bmod 50 = 19$.

21. $9 + 14 \equiv 4 + 4 \equiv 8 \equiv 3 \pmod 5$, so $(9 + 14) \bmod 5 = 3$.

23. $33 \cdot 96 \cdot 11 \equiv 3 \cdot 6 \cdot 1 \equiv 18 \equiv 8 \pmod{10}$, so $(33 \cdot 96 \cdot 11) \bmod 10 = 8$.

25. $15^3 - 25 \equiv 3^3 - 1 \equiv 27 - 1 \equiv 26 \equiv 2 \pmod 4$, so $(15^3 - 25) \bmod 4 = 2$.

27. $47(42 - 18) \equiv 2(6 - 0) \equiv 12 \equiv 3 \pmod 9$, so $(47(42 - 18)) \bmod 9 = 3$.

29. $8 \cdot 9 \cdot 10 \cdot 11 \cdot 12 \cdot 13 \equiv 1 \cdot 2 \cdot 3 \cdot 4 \cdot 5 \cdot 6 \equiv (1 \cdot 2 \cdot 3 \cdot 4)(5 \cdot 6) \equiv 24 \cdot 30 \equiv 3 \cdot 2 \equiv 6 \pmod 7$, so $(8 \cdot 9 \cdot 10 \cdot 11 \cdot 12 \cdot 13) \bmod 7 = 6$.

31. $2^{310} + 1 \equiv (-1)^{310} + 1 \equiv 1 + 1 \equiv 2 \pmod 3$, so $(2^{310} + 1) \bmod 3 = 2$.

33. $83 + 30 + 15 + 3 \equiv 2 + 3 + 6 + 3 \equiv 14 \equiv 5 \pmod 9$, so $(83 + 30 + 15 + 3) \bmod 9 = 5$.

35. $47 \cdot 50 \equiv 3 \cdot 6 \equiv 18 \pmod{44}$, so $(47 \cdot 50) \bmod 44 = 18$.

37. $42^5 + 17^{28} \equiv 2^5 + 1^{28} \equiv 32 + 1 \equiv 33 \equiv 1 \pmod 8$, so $(42^5 + 17^{28}) \bmod 8 = 1$.

39. $823 + 620 \equiv 23 + 20 \equiv 43 \pmod{100}$, so $(823 + 620) \bmod 100 = 43$.

41. Because $9^{2001} \equiv (-1)^{2001} \equiv -1 \equiv 9 \pmod{10}$, the last digit of 9^{2001} is 9.

43. Because $7453^{56} \equiv 3^{56} \equiv (3^2)^{28} \equiv 9^{28} \equiv (-1)^{28} \equiv 1 \pmod{10}$, the last digit of 7453^{56} is 1.

45. Let x = the number of soldiers. Because there are three soldiers left over when the general lines up his soldiers in 100 equal rows, we know $x \bmod 100 = 3$ and x can be written in the form $x = 100k + 3$, where k is a nonnegative integer. Because there are two soldiers left over when the general lines up his soldiers in 99 equal rows, we know $x \bmod 99 = 2$. Substituting $x = 100k + 3$ into the equation $x \bmod 99 = 2$, we have $(100k + 3) \bmod 99 = 2$. Using the properties of modular arithmetic, we solve for k and find

$$100k + 3 \equiv 2 \pmod{99}$$

$$1k + 3 \equiv 2 \pmod{99}$$

$$k \equiv 2 - 3 \pmod{99}$$

$$k \equiv -1 \pmod{99}$$

$$k \equiv -1 + 99 \equiv 98 \pmod{99}.$$

The only such nonnegative integer k yielding under 10,000 soldiers is $k = 98$. Therefore, the general has $100 \cdot 98 + 3 = 9803$ soldiers.

SECTION 8.3

1. Because $7 + 8 + 9 + 3 \equiv 7 + 8 + 3 \equiv 18 \equiv 0 \pmod 9$, 7893 is divisible by 9.

3. Because

$$1 + 9 + 5 + 7 + 1 + 9 + 5 + 8 + 1 + 9 + 8 + 6 + 1 + 9 + 8 + 9$$

$$\equiv (1 + 7 + 1) + 5 + 5 + (8 + 1) + (8 + 1) + 6 + 8 \equiv 5 + 5 + 6 + 8 \equiv 24 \equiv 6 \pmod 9,$$

1,957,195,819,861,989 is not divisible by 9.

5. Because

$$5 + 3 + 5 + 6 + 9 + 9 + 0 + 2 + 1 + 8 + 0 + 3 + 1 + 9 + 1 + 2 + 0 + 0 + 8$$
$$\equiv 5 + (3 + 6) + 5 + 2 + (1 + 8) + 3 + 1 + (1 + 8) + 2$$
$$\equiv 5 + 5 + 2 + 3 + 1 + 2 \equiv (5 + 3 + 1) + (5 + 2 + 2) \equiv 0 \pmod{9},$$

5,356,990,218,031,912,008 is divisible by 9.

7. Because $3 - 2 + 1 - 9 \equiv 3 + 1 - (2 + 9) \equiv 3 + 1 \equiv 4 \pmod{11}$, 9123 is not divisible by 11.

9. Because

$$2 - 0 + 9 - 0 + 5 - 0 + 1 - 4 + 7 \equiv (2 + 9) + 5 + 1 - 4 + 7 \equiv 5 + 1 - 4 + 7 \equiv 9 \pmod{11},$$

741,050,902 is not divisible by 11.

11. Because

$$1 - 2 + 3 - 4 + 5 - 6 + 7 - 8 + 9 - 9 + 8 - 7 + 6 - 5 + 4 - 3 + 2 - 1 \equiv 0 \pmod{11},$$

123,456,789,987,654,321 is divisible by 11.

13. Because 6 is not divisible by 4, it follows that 78,406 is not divisible by 4.

15. Because $84 = 4 \cdot 21$ is divisible by 4, it follows that 82,168,003,234 is divisible by 4.

17. (a) Its last digit is even, so 62,952 is divisible by 2.

(b) Because $6 + 2 + 9 + 5 + 2 \equiv 2 + 5 + 2 \equiv 9 \equiv 0 \pmod 3$, it follows that 62,952 is divisible by 3.

(c) Because $52 = 4 \cdot 13$ is divisible by 4, we see that 62,952 is divisible by 4.

(d) Its last digit is not 0 or 5, so 62,952 is not divisible by 5.

(e) We have shown that 62,952 is divisible by 2 and 3, and therefore it is also divisible by 6.

(f) Carrying out the test for divisibility by 7, we get

$$\begin{array}{ccccc} 6295 & & 629 & & 62 \\ \underline{-4} & \text{then} & \underline{-2} & \text{then} & \underline{-14} \\ 6291 & & 627 & & 48 \end{array}.$$

Because 48 is not divisible by 7, it follows that 62,952 is not divisible by 7.

(g) Because $952 = 8 \cdot 119$ is divisible by 8, we see that 62,952 is divisible by 8.

(h) Because $6 + 2 + 9 + 5 + 2 \equiv 6 + (2 + 5 + 2) \equiv 6 \pmod 9$, it follows that 62,952 is not divisible by 9.

(i) Its last digit is not 0, so 62,952 is not divisible by 10.

(j) Because $2 - 5 + 9 - 2 + 6 \equiv -5 + 9 + 6 \equiv 10 \pmod{11}$, we see that 62,952 is not divisible by 11.

19. (a) Its last digit is odd, so 2,756,768,175 is not divisible by 2.

(b) Because $2 + 7 + 5 + 6 + 7 + 6 + 8 + 1 + 7 + 5 \equiv 54 \equiv 0 \pmod 3$, it follows that 2,756,768,175 is divisible by 3.

(c) Because 75 is not divisible by 4, we see that 2,756,768,175 is not divisible by 4.

(d) Its last digit is 5, so 2,756,768,175 is divisible by 5.

(e) We have shown that 2,756,768,175 is not divisible by 2, so it is also not divisible by 6.

(f) Carrying out the test for divisibility by 7, we get

$$
\begin{array}{ccc}
\begin{array}{r} 275{,}676{,}817 \\ -10 \\ \hline 275{,}676{,}807 \end{array} & \text{then} &
\begin{array}{r} 27{,}567{,}680 \\ -14 \\ \hline 27{,}567{,}666 \end{array} & \text{then} &
\begin{array}{r} 2{,}756{,}766 \\ -12 \\ \hline 2{,}756{,}754 \end{array}
\end{array}
$$

$$
\text{then} \quad
\begin{array}{r} 275{,}675 \\ -8 \\ \hline 275{,}667 \end{array} \quad \text{then} \quad
\begin{array}{r} 27{,}566 \\ -14 \\ \hline 27{,}552 \end{array} \quad \text{then} \quad
\begin{array}{r} 2755 \\ -4 \\ \hline 2751 \end{array}
$$

$$
\text{then} \quad
\begin{array}{r} 275 \\ -2 \\ \hline 273 \end{array} \quad \text{then} \quad
\begin{array}{r} 27 \\ -6 \\ \hline 21 \end{array} \ .
$$

Because 21 is divisible by 7, it follows that 2,756,768,175 is divisible by 7.

(g) Because 175 is not divisible by 8, we see that 2,756,768,175 is not divisible by 8.

(h) Because

$$
2 + 7 + 5 + 6 + 7 + 6 + 8 + 1 + 7 + 5
$$
$$
\equiv (2 + 7) + 5 + 6 + 7 + 6 + (8 + 1) + 7 + 5
$$
$$
\equiv 5 + 6 + 7 + 6 + 7 + 5 \equiv 36 \equiv 0 \pmod{9},
$$

if follows that 2,756,768,175 is divisible by 9.

(i) Its last digit is not 0, so 2,756,768,175 is not divisible by 10.

(j) Because

$$
5 - 7 + 1 - 8 + 6 - 7 + 6 - 5 + 7 - 2
$$
$$
\equiv 1 - 8 + 6 - 7 + 6 - 2 \equiv -4 \equiv -4 + 11 \equiv 7 \pmod{11},
$$

it follows that 2,756,768,175 is not divisible by 11.

21. First note that $8 | 10^3$, so $10^k \equiv 0 \pmod{8}$ for $k \geq 3$. Let $N = a_m a_{m-1} \cdots a_1 a_0$ be a positive integer. Then

$$
N \equiv a_m \cdot 10^m + a_{m-1} \cdot 10^{m-1} + \cdots + a_3 \cdot 10^3 + a_2 \cdot 10^2 + a_1 \cdot 10 + a_0
$$
$$
\equiv a_m \cdot 0 + a_{m-1} \cdot 0 + \cdots + a_3 \cdot 0 + a_2 \cdot 10^2 + a_1 \cdot 10 + a_0
$$
$$
\equiv a_2 \cdot 10^2 + a_1 \cdot 10 + a_0 \pmod{8}.
$$

The number $a_2 \cdot 10^2 + a_1 \cdot 10 + a_0$ is just the three-digit number $a_2 a_1 a_0$ given by the last three digits of N. It follows that N and the number given by its last three digits are congruent modulo 8 and therefore leave the same remainder when divided by 8. Therefore, a positive integer is divisible by 8 if and only if the number formed by its last three digits is divisible by 8.

23. First note that $10 \equiv 1 \pmod{3}$. Let $N = a_m a_{m-1} \cdots a_1 a_0$ be a positive integer. Then

$$
N \equiv a_m \cdot 10^m + a_{m-1} \cdot 10^{m-1} + \cdots + a_1 \cdot 10 + a_0
$$
$$
\equiv a_m \cdot 1^m + a_{m-1} \cdot 1^{m-1} + \cdots + a_1 \cdot 1 + a_0
$$
$$
\equiv a_m + a_{m-1} + \cdots + a_1 + a_0 \pmod{3}.
$$

We see that N is congruent to the sum of its digits modulo 3, and therefore they leave the same remainder when divided by 3. Therefore, a positive integer is divisible by 3 if and only if the sum of its digits is 0 mod 3.

25. We know the number $4\#1$ is divisible by 9, and therefore $4 + \# + 1 \equiv 0 \pmod{9}$. It follows that $\#$ is 4, so the missing digit is 4. Each candy bar cost $\$4.41/9 = \$0.49$.

27. For any palindrome with an even number of digits, the alternating sum of the digits is zero due to cancellation. Therefore, it must be divisible by 11.

29. Using casting out nines, we find
$$7 + 3 + 0 + 9 + 6 + 3 + 7 + 0 + 9 \equiv 7 + (3 + 6) + 3 + 7 \equiv 17 \equiv 8 \pmod{9}$$
and
$$2 + 8 + 1 + 0 + 0 + 9 + 2 + 3 + 5 + 6 + 1 + 3 + 9 + 8$$
$$\equiv 2 + (8 + 1) + 2 + (3 + 6) + (5 + 1 + 3) + 8$$
$$\equiv 2 + 2 + 8 \equiv 12 \equiv 3 \pmod{9}.$$
Therefore,
$$730{,}963{,}709 \cdot 28{,}100{,}923{,}561{,}398 \equiv 8 \cdot 3 \equiv 24 \equiv 6 \pmod{9}.$$
Because
$$2 + 0 + 4 + 5 + 0 + 7 + 5 + 5 + 3 + 1 + 2 + 7 + 6 + 4 + 9 + 7 + 1 + 3 + 0 + 5 + 1 + 8 + 2$$
$$\equiv (2 + 7) + (4 + 5) + 5 + (5 + 3 + 1) + (2 + 7) + (6 + 3) + (4 + 5) + (7 + 2) + (1 + 8) + 1$$
$$\equiv 5 + 1 \equiv 6 \pmod{9},$$
it follows that the calculation could be correct.

31. Using casting out elevens, we find
$$9 - 0 + 7 - 3 + 6 - 9 + 0 - 3 + 7 \equiv 7 - 3 + 6 - 3 + 7 \equiv 14 \equiv 3 \pmod{11}$$
and
$$8 - 9 + 3 - 1 + 6 - 5 + 3 - 2 + 9 - 0 + 0 - 1 + 8 - 2$$
$$\equiv (8 + 3) - 1 + 6 - 5 + (3 + 8) - 2 - 1 - 2$$
$$\equiv -1 + 6 - 5 - 2 - 1 - 2 \equiv -5 \equiv -5 + 11 \equiv 6 \pmod{11}.$$
Therefore,
$$730{,}963{,}709 \cdot 28{,}100{,}923{,}561{,}398 \equiv 3 \cdot 6 \equiv 18 \equiv 7 \pmod{11}.$$
Because
$$2 - 8 + 1 - 5 + 0 - 3 + 1 - 7 + 9 - 4 + 6 - 7 + 2 - 1 + 3 - 5 + 5 - 7 + 0 - 5 + 4 - 0 + 2$$
$$\equiv (2 + 9) - 8 + 1 - 5 - 7 + 6 - 7 + 2 - 7 - 5 + 2$$
$$\equiv -8 + 1 - 5 - 7 + 6 - 7 + 2 - 7 - 5 + 2 \equiv -28 \equiv -28 + 3 \cdot 11 \equiv 5 \pmod{11},$$
it follows that the calculation could not be correct.

SECTION 8.4

1. Because $161311619298 \bmod 7 = 4$, the check digit is 4.

3. Because $798103588 \bmod 7 = 2$, the check digit will detect the error.

5. Because the error in recording the number was simply the transposition of the digits 51, the sum of the digits of the main part is the same for both the correct and incorrect numbers. It follows from casting out nines that the main parts of the incorrect and the correct numbers are the same modulo 9 (both are 6, but the calculation is not necessary in this case). Therefore, the check digit will not detect the error.

7. Using the bank identification number check digit formula, we find

$$a_0 \equiv 7 \cdot 1 + 3 \cdot 1 + 9 \cdot 1 + 7 \cdot 0 + 3 \cdot 0 + 9 \cdot 0 + 7 \cdot 0 + 3 \cdot 2 \equiv 7 + 3 + 9 + 6 \equiv 25 \equiv 5 \pmod{10}.$$

Therefore, the entire bank identification number is 111000025.

9. Using the bank identification number check digit formula, we get

$$a_0 \equiv 7 \cdot 1 + 3 \cdot 0 + 9 \cdot 0 + 7 \cdot 6 + 3 \cdot 4 + 9 \cdot 6 + 7 \cdot 0 + 3 \cdot 2 \equiv 7 + 42 + 12 + 54 + 6$$
$$\equiv 7 + 2 + 2 + 4 + 6 \equiv 21 \equiv 1 \not\equiv 3 \pmod{10},$$

so the check could not be authentic.

11. Using the UPC number check digit formula with the incorrectly scanned number, we have

$$a_0 \equiv -(3 \cdot 0 + 2 + 3 \cdot 8 + 0 + 3 \cdot 0 + 0 + 3 \cdot 2 + 9 + 3 \cdot 2 + 8 + 3 \cdot 0)$$
$$\equiv -(2 + 24 + 6 + 9 + 6 + 8) \equiv -(2 + 4 + 6 + 9 + 6 + 8)$$
$$\equiv -35 \equiv -35 + 4 \cdot 10 \equiv 5 \pmod{10},$$

so the scanner will expect to see 5 as the check digit.

13. Using the UPC number check digit formula with the incorrectly entered number, we have

$$a_0 \equiv -(3 \cdot 0 + 3 + 3 \cdot 9 + 8 + 3 \cdot 0 + 0 + 3 \cdot 0 + 3 + 3 \cdot 6 + 8 + 3 \cdot 7)$$
$$\equiv -(3 + 27 + 8 + 3 + 18 + 8 + 21) \equiv -(3 + 7 + 8 + 3 + 8 + 8 + 1)$$
$$\equiv -38 \equiv -38 + 4 \cdot 10 \equiv 2 \not\equiv 0 \pmod{10}.$$

Therefore, the check digit will detect the error.

15. Using the UPC number check digit formula, we have

$$a_0 \equiv -(3 \cdot 3 + 0 + 3 \cdot 0 + 8 + 3 \cdot 1 + 0 + 3 \cdot 8 + \# + 3 \cdot 0 + 2 + 3 \cdot 4)$$
$$\equiv -(9 + 8 + 3 + 24 + \# + 2 + 12) \equiv -(9 + 8 + 3 + 4 + \# + 2 + 2)$$
$$\equiv -(28 + \#) \equiv -(8 + \#) \equiv 9 \pmod{10}.$$

It follows that $8 + \# \equiv -9 \equiv -9 + 10 \equiv 1 \pmod{10}$, and there the digit $\#$ must be 3.

17. Using the ISBN check digit formula, we get

$$a_0 \equiv 0 + 2 \cdot 3 + 3 \cdot 9 + 4 \cdot 5 + 5 \cdot 3 + 6 \cdot 3 + 7 \cdot 9 + 8 \cdot 5 + 9 \cdot 7$$
$$\equiv 6 + 27 + 20 + 15 + 18 + 63 + 40 + 63 \equiv 6 + 5 + 9 + 4 + 7 + 8 + 7 + 8$$
$$\equiv 54 \equiv 10 \pmod{11}.$$

Therefore, the check digit is X.

19. Using the check digit formula, we find

$$a_0 \equiv -(4+1+7+9+4+8+9+1+9) \equiv -(4+1+7+4+8+1)$$
$$\equiv -(4+1+7+4+8+1) \equiv -25 \equiv -25 + 3 \cdot 9 \equiv 2 \ (\text{mod } 9).$$

Therefore, the 10-digit serial number is 4179489192.

21. **(a)** Using the check digit formula with the incorrectly recorded number, we have

$$a_0 \equiv -(2+6+7+0+1+1+3+0+9) \equiv -(2+6+7+1+1+3)$$
$$\equiv -20 \equiv -20 + 3 \cdot 9 \equiv 7 \ (\text{mod } 9),$$

so the check digit will detect the error.

(b) Replacing a 0 by a 9 or vice versa would not be detected by the check digit.

23. Using the check character formula, we find

$$a_0 \equiv -(15 \cdot 3 + 14 \cdot 4 + 13 \cdot 29 + 12 \cdot 1 + 11 \cdot 12 + 10 \cdot 27$$
$$+ 9 \cdot 9 + 8 \cdot 2 + 7 \cdot 4 + 6 \cdot 4 + 5 \cdot 34 + 4 \cdot 13 + 3 \cdot 3 + 2 \cdot 32)$$
$$\equiv -(45 + 56 + 377 + 12 + 132 + 270 + 81 + 16 + 28 + 24 + 170 + 52 + 9 + 64)$$
$$\equiv -1336 \equiv -1336 + 38 \cdot 36 \equiv 32 \ (\text{mod } 36).$$

Therefore, the check character is W.

25. Using the check character formula with the incorrect code we find

$$a_0 \equiv -(15 \cdot 29 + 14 \cdot 3 + 13 \cdot 1 + 12 \cdot 3 + 11 \cdot 17 + 10 \cdot 20 + 9 \cdot 2 + 8 \cdot 1$$
$$+ 7 \cdot 27 + 6 \cdot 7 + 5 \cdot 7 + 4 \cdot 2 + 3 \cdot 3 + 2 \cdot 34)$$
$$\equiv -(435 + 42 + 13 + 36 + 187 + 200 + 18 + 8 + 189 + 42 + 35 + 8 + 9 + 68)$$
$$\equiv -1290 \equiv -1290 + 36 \cdot 36 \equiv 6 \not\equiv 5 \ (\text{mod } 36).$$

Therefore, the check character will detect the error.

27. First note that $7 \equiv 7 - 10 \equiv -3 \ (\text{mod } 10)$, $3 \equiv 3 - 10 \equiv -7 \ (\text{mod } 10)$, and $9 \equiv 9 - 10 \equiv -1 \ (\text{mod } 10)$. Therefore, starting with the bank identification number check digit formula, we have

$$a_0 \equiv 7a_8 + 3a_7 + 9a_6 + 7a_5 + 3a_4 + 9a_3 + 7a_2 + 3a_1$$
$$\equiv -3a_8 - 7a_7 - a_6 - 3a_5 - 7a_4 - a_3 - 3a_2 - 7a_1$$
$$\equiv -(3a_8 + 7a_7 + a_6 + 3a_5 + 7a_4 + a_3 + 3a_2 + 7a_1) \ (\text{mod } 10).$$

Therefore, the check digit formula can be written alternately as

$$a_0 = -(3a_8 + 7a_7 + a_6 + 3a_5 + 7a_4 + a_3 + 3a_2 + 7a_1) \ \text{mod } 10.$$

SECTION 8.5

1.

Round

	1st	2nd	3rd	4th	5th	6th	7th	8th	9th
Team 1	9	Bye	2	3	4	5	6	7	8
Team 2	8	9	1	Bye	3	4	5	6	7
Team 3	7	8	9	1	2	Bye	4	5	6
Team 4	6	7	8	9	1	2	3	Bye	5
Team 5	Bye	6	7	8	9	1	2	3	4
Team 6	4	5	Bye	7	8	9	1	2	3
Team 7	3	4	5	6	Bye	8	9	1	2
Team 8	2	3	4	5	6	7	Bye	9	1
Team 9	1	2	3	4	5	6	7	8	Bye

3.

Round

	1st	2nd	3rd	4th	5th	6th	7th	8th	9th
Team 1	9	10	2	3	4	5	6	7	8
Team 2	8	9	1	10	3	4	5	6	7
Team 3	7	8	9	1	2	10	4	5	6
Team 4	6	7	8	9	1	2	3	10	5
Team 5	10	6	7	8	9	1	2	3	4
Team 6	4	5	10	7	8	9	1	2	3
Team 7	3	4	5	6	10	8	9	1	2
Team 8	2	3	4	5	6	7	10	9	1
Team 9	1	2	3	4	5	6	7	8	10
Team 10	5	1	6	2	7	3	8	4	9

5.

Round

	1st	2nd	3rd	4th	5th	6th	7th	8th	9th	10th	11th	12th	13th	14th	15th
Team 1	15	Bye	2	3	4	5	6	7	8	9	10	11	12	13	14
Team 2	14	15	1	Bye	3	4	5	6	7	8	9	10	11	12	13
Team 3	13	14	15	1	2	Bye	4	5	6	7	8	9	10	11	12
Team 4	12	13	14	15	1	2	3	Bye	5	6	7	8	9	10	11
Team 5	11	12	13	14	15	1	2	3	4	Bye	6	7	8	9	10
Team 6	10	11	12	13	14	15	1	2	3	4	5	Bye	7	8	9
Team 7	9	10	11	12	13	14	15	1	2	3	4	5	6	Bye	8
Team 8	Bye	9	10	11	12	13	14	15	1	2	3	4	5	6	7
Team 9	7	8	Bye	10	11	12	13	14	15	1	2	3	4	5	6
Team 10	6	7	8	9	Bye	11	12	13	14	15	1	2	3	4	5
Team 11	5	6	7	8	9	10	Bye	12	13	14	15	1	2	3	4
Team 12	4	5	6	7	8	9	10	11	Bye	13	14	15	1	2	3
Team 13	3	4	5	6	7	8	9	10	11	12	Bye	14	15	1	2
Team 14	2	3	4	5	6	7	8	9	10	11	12	13	Bye	15	1
Team 15	1	2	3	4	5	6	7	8	9	10	11	12	13	14	Bye

7.

	1st	2nd	3rd	4th	5th	6th	7th	8th	9th	10th	11th	12th	13th	14th	15th
Team 1	15	16	2	3	4	5	6	7	8	9	10	11	12	13	14
Team 2	14	15	1	16	3	4	5	6	7	8	9	10	11	12	13
Team 3	13	14	15	1	2	16	4	5	6	7	8	9	10	11	12
Team 4	12	13	14	15	1	2	3	16	5	6	7	8	9	10	11
Team 5	11	12	13	14	15	1	2	3	4	16	6	7	8	9	10
Team 6	10	11	12	13	14	15	1	2	3	4	5	16	7	8	9
Team 7	9	10	11	12	13	14	15	1	2	3	4	5	6	16	8
Team 8	16	9	10	11	12	13	14	15	1	2	3	4	5	6	7
Team 9	7	8	16	10	11	12	13	14	15	1	2	3	4	5	6
Team 10	6	7	8	9	16	11	12	13	14	15	1	2	3	4	5
Team 11	5	6	7	8	9	10	16	12	13	14	15	1	2	3	4
Team 12	4	5	6	7	8	9	10	11	16	13	14	15	1	2	3
Team 13	3	4	5	6	7	8	9	10	11	12	16	14	15	1	2
Team 14	2	3	4	5	6	7	8	9	10	11	12	13	16	15	1
Team 15	1	2	3	4	5	6	7	8	9	10	11	12	13	14	16
Team 16	8	1	9	2	10	3	11	4	12	5	13	6	14	7	15

(Round)

9. Because $T_{9,7} \equiv 7 - 9 \equiv -2 \equiv -2 + 19 \equiv 17 \,(\text{mod } 19)$, Team 17 will be assigned to play Team 9 in round 7.

11. Because $T_{22,12} \equiv 12 - 22 \equiv -10 \equiv -10 + 31 \equiv 21 \,(\text{mod } 31)$, Team 21 will be assigned to play Team 22 in round 12.

13. We have $T_{7,r} \equiv r - 7 \equiv 10 \,(\text{mod } 21)$, and therefore $r \equiv 10 + 7 \equiv 17 \,(\text{mod } 21)$. It follows that Team 7 will play Team 10 in round 17.

15. We have $T_{12,r} \equiv r - 12 \equiv 12 \,(\text{mod } 17)$, and therefore $r \equiv 12 + 12 \equiv 24 \equiv 7 \,(\text{mod } 17)$. It follows that Team 12 will have a bye in round 7.

17. For the schedule, refer back to Exercise 3.

Home teams: Round 1: 1, 2, 3, 4, 10; Round 2: 6, 7, 8, 9, 10; Round 3: 2, 3, 4, 5, 6; Round 4: 1, 2, 7, 8, 9; Round 5: 3, 4, 5, 6, 10; Round 6: 1, 2, 8, 9, 10; Round 7: 4, 5, 6, 7, 8; Round 8: 1, 2, 3, 4, 9; Round 9: 5, 6, 7, 8, 10

19. For the schedule, refer back to Exercise 5.

Home teams: Round 1: 1, 2, 3, 4, 5, 6, 7; Round 2: 9, 10, 11, 12, 13, 14, 15; Round 3: 2, 3, 4, 5, 6, 7, 8; Round 4: 1, 10, 11, 12, 13, 14, 15; Round 5: 3, 4, 5, 6, 7, 8, 9; Round 6: 1, 2, 11, 12, 13, 14, 15; Round 7: 4, 5, 6, 7, 8, 9, 10; Round 8: 1, 2, 3, 12, 13, 14, 15; Round 9: 5, 6, 7, 8, 9, 10, 11; Round 10: 1, 2, 3, 4, 13, 14, 15; Round 11: 6, 7, 8, 9, 10, 11, 12; Round 12: 1, 2, 3, 4, 5, 14, 15; Round 13: 7, 8, 9, 10, 11, 12, 13; Round 14: 1, 2, 3, 4, 5, 6, 15; Round 15: 8, 9, 10, 11, 12, 13, 14

21. For i odd, Team i will be the home team versus Team j when j is even and $j < i$ and when j is odd and $j > i$. Thus, when i is odd, Team i will be the home team against Teams $2, 4, \ldots, i-1$ and Teams $i+2, i+4, \ldots, N$, for a total of

$$\frac{i-1}{2} + \frac{N-i}{2} = \frac{N-1}{2}$$

home games. Similarly, for i even, Team i will be the home team against Teams $1, 3, \ldots, i-1$ and Teams $i+2, i+4, \ldots, N-1$, for a total of

$$\frac{i}{2} + \frac{N-1-i}{2} = \frac{N-1}{2}$$

home games.

23. For the schedule, refer back to Exercise 3.

    Home teams: Round 1: 3, 4, 5, 8, 9; Round 2: 3, 4, 5, 9, 10; Round 3: 1, 4, 5, 6, 9; Round 4: 1, 4, 5, 6, 10; Round 5: 1, 2, 5, 6, 7; Round 6: 1, 2, 6, 7, 10; Round 7: 1, 2, 3, 7, 8; Round 8: 2, 3, 7, 8, 10; Round 9: 2, 3, 4, 8, 9

25. For the schedule, refer back to Exercise 5.

    Home teams: Round 1: 5, 6, 7, 12, 13, 14, 15; Round 2: 5, 6, 7, 8, 13, 14, 15; Round 3: 1, 6, 7, 8, 13, 14, 15; Round 4: 1, 6, 7, 8, 9, 14, 15; Round 5: 1, 2, 7, 8, 9, 14, 15; Round 6: 1, 2, 7, 8, 9, 10, 15; Round 7: 1, 2, 3, 8, 9, 10, 15; Round 8: 1, 2, 3, 8, 9, 10, 11; Round 9: 1, 2, 3, 4, 9, 10, 11; Round 10: 2, 3, 4, 9, 10, 11, 12; Round 11: 2, 3, 4, 5, 10, 11, 12; Round 12: 3, 4, 5, 10, 11, 12, 13; Round 13: 3, 4, 5, 6, 11, 12, 13; Round 14: 4, 5, 6, 11, 12, 13, 14; Round 15: 4, 5, 6, 7, 12, 13, 14

27. Note that when N is odd, $N/2$ is not an integer. First consider $k < N/2$. Then Team k will be the home team against Teams $k+1, k+2, \ldots, k+(N-1)/2$ and will be the away team otherwise, for a total of $(N-1)/2$ home games. When $k > N/2$, Team k will be the away team against Teams $k-(N-1)/2, k-(N-1)/2+1, \ldots, k-1$ and will be the home team otherwise, for a total of $(N-1)-(N-1)/2 = (N-1)/2$ home games.

29. Using the given formula, we see that in round 1 Team $T_{1,1}$ is assigned to play Team 1, where $T_{1,1} \equiv 2 \cdot 1 - 1 \equiv 1 \,(\text{mod } 7)$. Therefore, Team 1 is given a bye in round 1. Similarly, in round 1 Team $T_{2,1}$ is assigned to play Team 2, where $T_{2,1} \equiv 2 \cdot 1 - 2 \equiv 0 \equiv 7 \,(\text{mod } 7)$, and therefore Team 7 is assigned to play Team 2. Continuing in this way, we arrive at the following schedule:

Round

	1st	2nd	3rd	4th	5th	6th	7th
Team 1	Bye	3	5	7	2	4	6
Team 2	7	Bye	4	6	1	3	5
Team 3	6	1	Bye	5	7	2	4
Team 4	5	7	2	Bye	6	1	3
Team 5	4	6	1	3	Bye	7	2
Team 6	3	5	7	2	4	Bye	1
Team 7	2	4	6	1	3	5	Bye

The only difference between this schedule and the one arrived at in Example 1 is that the rounds have been scrambled.

31. Suppose Team j is assigned to play both Team m and Team k in round r. Then $j \equiv r - m \,(\text{mod } N)$ and $j \equiv r - k \,(\text{mod } N)$, implying $r - m \equiv r - k \,(\text{mod } N)$. Therefore, $k \equiv m \,(\text{mod } N)$. Because $1 \le m \le N$ and $1 \le k \le N$, it follows that $k = m$.

33. Team m is assigned to play itself in round r if and only if $m \equiv r - m \,(\text{mod } N)$ or $2m \equiv r \,(\text{mod } N)$. First note that $2m$ is a positive even number between 2 and $2N$. In order also to be congruent to r modulo N, $2m$ must equal r or $N + r$. However, because N is odd, exactly one of these two values is even, so there is exactly one such Team m.

SECTION 8.6

1. First, we break the message into blocks and get

ANYTI MEISO KAY.

Converting to their numerical equivalents, we get

$$0 \ 13 \ 24 \ 19 \ 8 \qquad 12 \ 4 \ 8 \ 18 \ 14 \qquad 10 \ 0 \ 24.$$

Enciphering the first number, 0, we get $C \equiv 0 + 3 \equiv 3 \, (\text{mod } 26)$. We encipher the remaining numbers similarly and get

$$3 \ 16 \ 1 \ 22 \ 11 \qquad 15 \ 7 \ 11 \ 21 \ 17 \qquad 13 \ 3 \ 1.$$

Converting to their letter equivalents, we get the enciphered message

DQBWL PHLVR NDB.

3. Converting the letters in the enciphered message into their numerical equivalents, we get

$$1 \ 17 \ 23 \ 9 \ 17 \qquad 22 \ 11 \ 22 \ 20 \ 11 \qquad 9 \ 10 \ 22.$$

Deciphering the first number, 1, we get $P \equiv 1 - 3 \equiv -2 \equiv 24 \, (\text{mod } 26)$. We decipher the remaining numbers similarly and get

$$24 \ 14 \ 20 \ 6 \ 14 \qquad 19 \ 8 \ 19 \ 17 \ 8 \qquad 6 \ 7 \ 19.$$

Converting to their letter equivalents, we have

YOUGO TITRI GHT.

We see that the deciphered message is YOU GOT IT RIGHT.

5. First, we break the message into blocks and get

YOUSH OULDS TAY.

Converting to their numerical equivalents, we get

$$24 \ 14 \ 20 \ 18 \ 7 \qquad 14 \ 20 \ 11 \ 3 \ 18 \qquad 19 \ 0 \ 24.$$

Enciphering the first number, 24, we get $C \equiv 5 \cdot 24 + 14 \equiv 134 \equiv 4 \, (\text{mod } 26)$. We encipher the remaining numbers similarly and get

$$4 \ 6 \ 10 \ 0 \ 23 \qquad 6 \ 10 \ 17 \ 3 \ 0 \qquad 5 \ 14 \ 4.$$

Converting to their letter equivalents, we get the enciphered message

EGKAX GKRDA FOE.

7. First, we break the message into blocks and get

HEREI STHES ECRET.

Converting to their numerical equivalents, we get

$$7 \ 4 \ 17 \ 4 \ 8 \qquad 18 \ 19 \ 7 \ 4 \ 18 \qquad 4 \ 2 \ 17 \ 4 \ 19.$$

Enciphering the first number, 7, we get $C \equiv 19 \cdot 7 + 7 \equiv 140 \equiv 10 \, (\text{mod } 26)$. We encipher the remaining numbers similarly and get

$$10 \ 5 \ 18 \ 5 \ 3 \qquad 11 \ 4 \ 10 \ 5 \ 11 \qquad 5 \ 19 \ 18 \ 5 \ 4.$$

Converting to their letter equivalents, we get the enciphered message

KFSFD LEKFL FTSFE.

9. Converting the letters in the enciphered message into their numerical equivalents, we get

$$17\ 15\ 16\ 2\ 20 \qquad 19\ 16\ 6\ 16\ 20 \qquad 3.$$

The deciphering formula is $P = (15(C-16))\bmod 26$. Deciphering the first number, 17, we get $P \equiv 15(17-16) \equiv 15 \cdot 1 \equiv 15\,(\bmod\,26)$. We decipher the remaining numbers similarly and get

$$15\ 11\ 0\ 24\ 8 \qquad 19\ 0\ 6\ 0\ 8 \qquad 13.$$

Converting to their letter equivalents, we have

PLAYI TAGAIN.

We see that the deciphered message is PLAY IT AGAIN.

11. Converting the letters in the enciphered message into their numerical equivalents, we get

$$0\ 19\ 22\ 4\ 4 \qquad 14\ 11\ 22\ 5\ 4 \qquad 21\ 11\ 3\ 23.$$

The deciphering formula is $P = (17(C-9))\bmod 26$. Deciphering the first number, 0, we get $P \equiv 17(0-9) \equiv 17 \cdot (-9) \equiv -153 \equiv 3\,(\bmod\,26)$. We decipher the remaining numbers similarly and get

$$3\ 14\ 13\ 19\ 19 \qquad 7\ 8\ 13\ 10\ 19 \qquad 22\ 8\ 2\ 4.$$

Converting to their letter equivalents, we have

DONTT HINKT WICE.

We see that the deciphered message is DON'T THINK TWICE.

13. **(a)** Substituting 6, the numerical equivalent of G, into the enciphering formula, we get $P \equiv 6 \cdot 6 + 5 \equiv 41 \equiv 15\,(\bmod\,26)$. The letter equivalent of 15 is P, so G is encoded as P.

   (b) Substituting 19, the numerical equivalent of T, into the enciphering formula, we get $P \equiv 6 \cdot 19 + 5 \equiv 119 \equiv 15\,(\bmod\,26)$. Therefore, T is also encoded as P.

   (c) When a P appears in an encoded message, there is no way to tell if it is an encoded G or an encoded P, so the message can not be properly decoded.

15. We can encode the digits 0 through 9, the 26 letters, and blank spaces by using the encoding formula $C = (aP+b)\bmod 37$, where a and b are both integers between 0 and 36, and $\gcd(a, 37) = 1$. (Because 37 is prime, a can be any number between 1 and 36.) In this case, we let the letter A be represented by 10, B by 11, up to letting Z be represented by 36, and we let a blank space be represented by 37.

SECTION 8.7

1. First, we break the message into two-letter blocks and get

AS TO RM IS CO MI NG.

Converting to their numerical equivalents, we get

$$0\ 18 \qquad 19\ 14 \qquad 17\ 12 \qquad 8\ 18 \qquad 2\ 14 \qquad 12\ 8 \qquad 13\ 6.$$

Enciphering the first block, 0 18, we have

$$C_1 \equiv 4 \cdot 0 + 3 \cdot 18 \equiv 54 \equiv 2\,(\bmod\,26),$$
$$C_2 \equiv 15 \cdot 0 + 5 \cdot 18 \equiv 90 \equiv 12\,(\bmod\,26),$$

so it is encoded as the block 2 12. We encode the remaining blocks similarly and get

$$2\ 12 \qquad 14\ 17 \qquad 0\ 3 \qquad 8\ 2 \qquad 24\ 22 \qquad 20\ 12 \qquad 18\ 17.$$

Converting to their letter equivalents, we get the encoded message

CM OR AD IC YW UM SR.

3. First, we break the message into two-letter blocks and get

GO FO RI TX.

Converting to their numerical equivalents, we get

6 14   5 14   17 8   19 23.

Enciphering the first block, 6 14, we have

$$C_1 \equiv 5 \cdot 6 + 1 \cdot 14 \equiv 44 \equiv 18 \pmod{26},$$
$$C_2 \equiv 6 \cdot 6 + 13 \cdot 14 \equiv 218 \equiv 10 \pmod{26},$$

so it is encoded as the block 18 10. We encode the remaining blocks similarly and get

18 10   13 4   15 24   14 23.

Converting to their letter equivalents, we get the encoded message

SK NE PY OX.

5. Converting the letters in the enciphered message into their numerical equivalents, we get

3 5   25 7   19 25   15 9   20 8.

To find the deciphering formula note that $ad - bc \equiv 5 \cdot 7 - 2 \cdot 3 \equiv 29 \equiv 3 \pmod{26}$ and $\overline{ad - bc}$ mod $26 = \overline{3}$ mod $26 = 9$. Therefore,

$$A \equiv 9 \cdot 7 \equiv 63 \equiv 11 \pmod{26}, \qquad B \equiv -9 \cdot 2 \equiv -18 \equiv 8 \pmod{26},$$
$$C \equiv -9 \cdot 3 \equiv -27 \equiv 25 \pmod{26}, \qquad D \equiv 9 \cdot 5 \equiv 45 \equiv 19 \pmod{26},$$

and the deciphering formulas are given by

$$P_1 = (11C_1 + 8C_2) \mod 26$$
$$P_2 = (25C_1 + 19C_2) \mod 26.$$

Translating the first block, 3   5, we get

$$P_1 \equiv 11 \cdot 3 + 8 \cdot 5 \equiv 73 \equiv 21 \pmod{26},$$
$$P_2 \equiv 25 \cdot 3 + 19 \cdot 5 \equiv 170 \equiv 14 \pmod{26},$$

and therefore the first block translates into the block 21   14. Similarly, we translate each of the remaining blocks, and the entire message is translated to

21 14   19 4   19 14   3 0   24 2.

Converting to their letter equivalents, we have

VO TE TO DA YC.

We see that the decoded message is VOTE TODAY.

7. Converting the letters in the enciphered message into their numerical equivalents, we get

20 1   13 2   18 15   18 1   21 9.

To find the deciphering formula note that $ad - bc \equiv 6 \cdot 3 - 7 \cdot 19 \equiv -115 \equiv 15 \pmod{26}$ and $\overline{ad - bc} \bmod 26 = \overline{15} \bmod 26 = 7$. Therefore,

$$A \equiv 7 \cdot 3 \equiv 21 \pmod{26}, \qquad\qquad B \equiv -7 \cdot 7 \equiv -49 \equiv 3 \pmod{26},$$

$$C \equiv -7 \cdot 19 \equiv -133 \equiv 23 \pmod{26}, \qquad D \equiv 7 \cdot 6 \equiv 42 \equiv 16 \pmod{26},$$

and the deciphering formulas are given by

$$P_1 = (21C_1 + 3C_2) \bmod 26$$
$$P_2 = (23C_1 + 16C_2) \bmod 26.$$

Translating the first block, 20 1, we get

$$P_1 \equiv 21 \cdot 20 + 3 \cdot 1 \equiv 423 \equiv 7 \pmod{26},$$
$$P_2 \equiv 23 \cdot 20 + 16 \cdot 1 \equiv 476 \equiv 8 \pmod{26},$$

and therefore the first block translates into the block 7 8. Similarly, we translate each of the remaining blocks, and the entire message is translated to

$$7 \ 8 \qquad 19 \ 19 \qquad 7 \ 4 \qquad 17 \ 14 \qquad 0 \ 3.$$

Converting to their letter equivalents, we have

$$\text{HI} \quad \text{TT} \quad \text{HE} \quad \text{RO} \quad \text{AD.}$$

We see that the decoded message is HIT THE ROAD.

9. Translating all the letters into two-digit numbers, we get

$$19 \ 04 \ 11 \ 11 \qquad 12 \ 04 \qquad 22 \ 07 \ 24.$$

Forming blocks of length three, we get

$$190 \qquad 411 \qquad 111 \qquad 204 \qquad 220 \qquad 724.$$

Using the encryption formula $C = P^3 \bmod 3071$ to encode the first block, $P = 190$, we get $C \equiv 190^3 \equiv 6{,}859{,}000 \equiv 1457 \pmod{3071}$, so the block 190 is encoded as 1457. Continuing in this way, we encode the remaining blocks and the entire message is encoded as

$$1457 \qquad 0434 \qquad 1036 \qquad 1420 \qquad 0843 \qquad 1528.$$

11. Translating all the letters into two-digit numbers, we get

$$08 \qquad 05 \ 14 \ 20 \ 13 \ 03 \qquad 08 \ 19.$$

Forming blocks of length three, we get

$$080 \qquad 514 \qquad 201 \qquad 303 \qquad 081 \qquad 900.$$

Using the encryption formula $C = P^5 \bmod 7081$ to encode the first block, $P = 080$, we get

$$C \equiv 80^5 \equiv 80^3 \cdot 80^2 \equiv 512{,}000 \cdot 6400$$
$$\equiv 2168 \cdot 6400 \equiv 13{,}875{,}200 \equiv 3521 \pmod{7081},$$

so the block 080 is encoded as 3521. Continuing in this way, we encode the remaining blocks and the entire message is encoded as

$$3521 \qquad 4185 \qquad 3906 \qquad 4101 \qquad 0867 \qquad 2704.$$

13. We use the decoding formula $P = C^3 \bmod 1711$. Translating the first block, 1472, we get

$$P \equiv 1472^3 \equiv 1472^2 \cdot 1472 \equiv 2{,}166{,}784 \cdot 1472$$

$$\equiv 658 \cdot 1472 \equiv 968{,}576 \equiv 150 \ (\bmod \ 1711),$$

and therefore the first block translates into the block 150. Similarly, we translate each of the remaining four-digit blocks, and the entire message is translated to

$$150 \quad 017 \quad 031 \quad 413 \quad 120 \quad 400.$$

Breaking them into the two-digit blocks representing the plaintext letters, we have

$$15 \quad 00 \quad 17 \quad 03 \quad 14 \quad 13 \quad 12 \quad 04 \quad 00,$$

and translating these numbers to their letter equivalents we have

$$\text{P A R D O N M E A},$$

so we can see that the original message was PARDON ME.

15. We use the decoding formula $P = C^7 \bmod 4387$. Translating the first block, 3542, we get

$$P \equiv 3542^7 \equiv 3542^2 \cdot 3542^2 \cdot 3542^2 \cdot 3542$$

$$\equiv 12{,}545{,}764 \cdot 12{,}545{,}764 \cdot 12{,}545{,}764 \cdot 3542$$

$$\equiv (3331 \cdot 3331) \cdot (3331 \cdot 3542) \equiv 11{,}095{,}561 \cdot 11{,}798{,}402$$

$$\equiv 838 \cdot 1759 \equiv 1{,}474{,}042 \equiv 10 \ (\bmod \ 4387),$$

and therefore the first block translates into the block 010. Similarly, we translate each of the remaining four-digit blocks, and the entire message is translated to

$$010 \quad 418 \quad 201 \quad 704.$$

Breaking them into the two-digit blocks representing the plaintext letters, we have

$$01 \quad 04 \quad 18 \quad 20 \quad 17 \quad 04,$$

and translating these numbers to their letter equivalents we have

$$\text{B E S U R E},$$

so we can see that the original message was BE SURE.

17. **(a)** Converting SH to its numerical equivalent, we get 18 7. Enciphering 18 7, we have

$$C_1 \equiv 5 \cdot 18 + 3 \cdot 7 \equiv 111 \equiv 7 \ (\bmod \ 26),$$

$$C_2 \equiv 2 \cdot 18 + 9 \cdot 7 \equiv 99 \equiv 21 \ (\bmod \ 26),$$

so it is encoded as the block 7 21. Translating to letters, we see that SH is encoded as HV.

(b) Converting CZ to its numerical equivalent, we get 2 25. Enciphering 2 25, we have

$$C_1 \equiv 5 \cdot 2 + 3 \cdot 25 \equiv 85 \equiv 7 \ (\bmod \ 26),$$

$$C_2 \equiv 2 \cdot 2 + 9 \cdot 25 \equiv 229 \equiv 21 \ (\bmod \ 26),$$

so it is encoded as the block 7 21. Translating to letters, we see that CZ is encoded as HV.

(c) When the block HV appears in an encoded message, there is no way to tell if it is an encoded SH or an encoded CZ, so the message cannot be properly decoded.

19. The secret exponent k could be used for encoding messages and the public exponent r could be used to decode them.

Chapter 8 Review Exercises

1. $4312 = 2^3 \cdot 7^2 \cdot 11$

3. **(a)** $q = 6, r = 11$

   (b) $q = -17, r = 4$

5. $7 + 15 + 20 \equiv 7 + 6 + 2 \equiv 15 \equiv 6 \pmod 9$, so $(7 + 15 + 20) \bmod 9 = 6$.

7. $15^{300} + 13^{299} \equiv 1^{300} + (-1)^{299} \equiv 1 - 1 \equiv 0 \pmod 7$, so $(15^{300} + 13^{299}) \bmod 7 = 0$.

9. Because

$$4 + 2 + 0 + 9 + 8 + 7 + 9 + 3 + 3 + 9 + 0 + 1 + 1 + 5 + 8 + 6 + 6 + 0 + 4 + 5$$
$$\equiv 4 + (2 + 7) + 8 + 3 + 3 + 1 + (1 + 8) + 5 + 6 + 6 + (4 + 5)$$
$$\equiv 4 + 8 + 3 + 3 + 1 + 5 + 6 + 6 \equiv 36 \equiv 0 \pmod 9,$$

42,098,793,390,115,866,045 is divisible by 9.

11. Because $28 = 4 \cdot 7$ is divisible by 4, it follows that 47,877,905,008,564,228 is divisible by 4.

13. We know the number 20#9 is divisible by 11, and therefore $9 - \# + 0 - 2 \equiv 7 - \# \equiv 0 \pmod{11}$. It follows that the missing digit is 7, so the price for the eleven bags of potato chips was \$20.79. Therefore, each bag of chips cost \$20.79/11 = \$1.89.

15. Because $2780143911 \bmod 7 = 6$, the check digit is 6.

17. Because

$$4 + 0 + 9 + 8 + 3 + 2 + 9 + 6 + 5 + 8 \equiv (4 + 5) + 8 + (3 + 6) + 2 + 8$$
$$\equiv 8 + 2 + 8 \equiv 18 \equiv 0 \not\equiv 7 \pmod 9,$$

the check digit will detect the error.

19. Using the UPC number check digit formula, we have

$$a_0 \equiv -(3 \cdot 0 + 7 + 3 \cdot \# + 6 + 3 \cdot 2 + 8 + 3 \cdot 0 + 7 + 3 \cdot 1 + 5 + 3 \cdot 0)$$
$$\equiv -(7 + 3 \cdot \# + 6 + 6 + 8 + 7 + 3 + 5) \equiv -(42 + 3 \cdot \#)$$
$$\equiv -(2 + 3 \cdot \#) \equiv 4 \pmod{10}.$$

Therefore, $2 + 3 \cdot \# \equiv -4 \equiv -4 + 10 \equiv 6 \pmod{10}$. Trying the values $0, 1, 2, \ldots, 9$ for #, we find that the missing digit # is 8.

21. Because $T_{11,6} \equiv 6 - 11 \equiv -5 \equiv -5 + 19 \equiv 14 \pmod{19}$, Team 11 will play Team 14 in round 6.

23. First, we break the message into blocks and get

BRING SOMEM ONEY.

Converting to their numerical equivalents, we get

$$1 \ 17 \ 8 \ 13 \ 6 \qquad 18 \ 14 \ 12 \ 4 \ 12 \qquad 14 \ 13 \ 4 \ 24.$$

Enciphering the first number, 1, we get $C \equiv 7 \cdot 1 + 20 \equiv 27 \equiv 1 \, (\text{mod } 26)$. We encipher the remaining numbers similarly and get

$$1 \ 9 \ 24 \ 7 \ 10 \qquad 16 \ 14 \ 0 \ 22 \ 0 \qquad 14 \ 7 \ 22 \ 6.$$

Converting to their letter equivalents, we get the enciphered message

$$\text{BJYHK} \quad \text{QOAWA} \quad \text{OHWG.}$$

25. First, we break the message into two-letter blocks and get

$$\text{TU} \ \text{RN} \ \text{LE} \ \text{FT.}$$

Converting to their numerical equivalents, we get

$$19 \ 20 \qquad 17 \ 13 \qquad 11 \ 4 \qquad 5 \ 19.$$

Enciphering the block, 19 20, we have

$$C_1 \equiv 9 \cdot 19 + 4 \cdot 20 \equiv 251 \equiv 17 \ (\text{mod } 26),$$
$$C_2 \equiv 5 \cdot 19 + 23 \cdot 20 \equiv 555 \equiv 9 \ (\text{mod } 26),$$

so it is encoded as the block 17 9. We encode the remaining blocks similarly and get

$$17 \ 9 \qquad 23 \ 20 \qquad 11 \ 17 \qquad 17 \ 20.$$

Converting to their letter equivalents, we get the encoded message

$$\text{RJ} \ \text{XU} \ \text{LR} \ \text{RU.}$$

27. Translating all the letters into two-digit numbers, we get

$$01 \ 04 \ 06 \ 08 \ 13.$$

Forming blocks of length three, we get

$$010 \qquad 406 \qquad 081 \qquad 300.$$

Using the encryption formula $C = P^3 \bmod 4189$ to encode the first block, $P = 010$, we get $C \equiv 10^3 \equiv 1000 \, (\text{mod } 4189)$, so the block 010 is encoded as 1000. Continuing in this way, we encode the remaining blocks and the entire message is encoded as

$$1000 \qquad 4141 \qquad 3627 \qquad 1895.$$

CHAPTER 9
Student Solution Manual

SECTION 9.1

1. **(a)** If Player A chooses a first, he or she can win by responding to c or d with h or i, respectively.

   (b)

   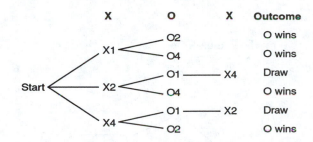

3. **(a)** The initial position of this game is

1	2	X
4	O	O
X	O	X

   The game tree is

   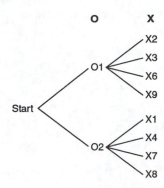

   (b) Player O should win under best play. No matter what Player X chooses, Player O has a winning response, either O2 or O4.

5. Player O can respond with a corner, which we represent by O1, or with a side, which we represent by O2. After O1, Player X can respond with X2 or equivalently X4, with X3 or equivalently X7, with X6 or equivalently X8, or with X9. Similarly after O2, Player X can choose X1 or equivalently X3, with X4 or equivalently X6, with X7 or equivalently X9, or with X8. The game tree is

7. The board looks like the following.

X	2	3
4	O	X
7	8	9

One way to show that Player O can achieve at least a draw with optimal play is with a partial game tree, showing only O's optimal responses. One possibility is the following

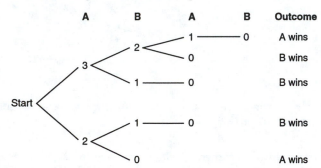

	O	X	O	X	O	X	Outcome
		X3	O8				O wins
		X4	O8				O wins
Start	O2	X7	O8				O wins
				X3	O9	X4	Draw
		X8	O7	X4	O3		O wins
				X9	O3		O wins
		X9	O8				O wins

9. (a) Denote a move by the number of pebbles left in the pile.

	A	B	A	B	Outcome
			1	0	A wins
	3	2			
			0		B wins
		1	0		B wins
Start					
	2	1	0		B wins
		0			A wins

(b) B wins under best play by choosing to leave 1 pebble with his or her first move.

11. (a) Write a move by the number of pebbles left in the two piles, compressing the game tree by putting the larger pile first.

	A	B	A	B	Outcome
		(1, 0)	(0, 0)		A wins
	(2, 0)				
		(0, 0)			B wins
		(1, 0)	(0, 0)		B wins
Start	(2, 0)				
		(0, 0)			A wins
	(2, 1)	(1, 1)	(1, 0)	(0, 0)	B wins
		(1, 0)	(0, 0)		A wins

(b) B wins under best play, responding (0, 0) to (2, 0) and (1, 1) to (2, 1).

13. **(a)** The player going first can win for exactly those n not divisible by 5. This player can remove one, two, three, or four pebbles to create a pile with number of pebbles a multiple of 5. The second player cannot leave a multiple of 5 pebbles in the pile. In particular, this player cannot remove the last pebble. Analogously, the player to go second wins under best play when n is divisible by 5.

  (b) Because one million is a multiple of 5, the second player wins under best play by always leaving a multiple of 5 pebbles in the pile at the end of his/her turn.

15. The safe points are $(1, 0)$, $(0, 1)$ and (n, n), with $n > 1$. Clearly $(1, 0)$ and $(0, 1)$ are safe because the next move takes the last pebble. A single pile with more than one pebble is not safe because a player can remove all but one, leaving a safe position. Nor is it safe when one pile has one pebble and the second at least one, because a player can remove all of the second pile. When both piles have n pebbles with $n > 1$, the next player to go must remove pebbles so that what remains is a single pile with more than one pebble, a pile with one pebble and a pile with more than one pebble, or two unequal piles of at least two pebbles. We have seen that the first two cases are not safe. In the third case, the player can remove pebbles from the larger pile so that they have the same number. It follows that two piles of n pebbles with $n > 1$ is safe and that two unequal piles of at least two pebbles is not safe.

17. **(a)**

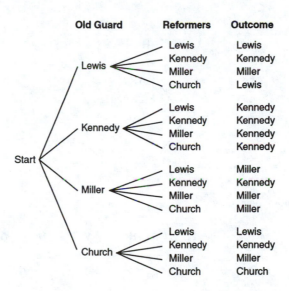

  (b) We show one possibility for the Reformers' responses under best play to each of the Old Guard's nominations in a partial game tree.

  We see that the Old Guard should nominate Lewis, and the Reformers should nominate Lewis in response. Lewis will be elected.

19. Right wins under optimal play. One possible partial game tree illustrating this strategy follows.

```
         L  ·R   L   R   L   R   Outcome
        4 — 2 — 6 — 5 — 3 — 7   Right wins
Start <
        6 — 2 — 4 — 5 — 3 — 7   Right wins
```

21. Left wins under optimal play. The partial game tree illustrating this strategy follows.

```
          L   R   L   R   L   R   L   R   L   Outcome
                      2 — 1            Left wins
Start — 3 — 4 — 5 <           3 — 1   Left wins
                      7 — 2 <
                              6 — 4 — 3 — 1   Left wins
```

23. (a)

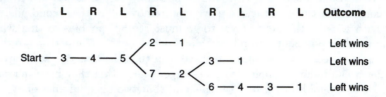

(b) H wins under best play.

```
         H       V       H     Outcome
Start — 2, 3 —— 1, 5 —— 7, 8    H wins
```

25. **(a)**

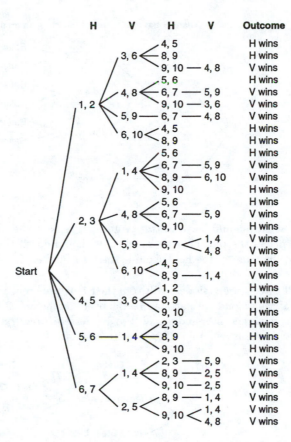

	H	V	H	V	Outcome

(tree) — see figure

(b) H wins under best play. One possible partial compressed game tree illustrating a winning strategy follows.

	H	V	H	Outcome
Start —	4, 5 —	3, 6 —	1, 2	H wins

27. H wins by placing a tile in the middle row if n is odd and one of the middle two rows if n is even, say at the start of the row. This blocks the first k columns from V, and V can place only one tile in each of the last $n - k$ columns. On the other hand, H can play a tile at the start of every row without hindrance, another $n - 1$ tiles. Thus V will be unable to move before H is unable to move.

29. **(a)**

Offense	Defense	Outcome
Regular	Regular	40%
	Goal line	50%
Goal line	Regular	70%
	Goal line	60%

Start

(b) The defense will respond to the regular and goal line offenses with its regular and goal line defenses, respectively, for offensive success rates of 40% and 60%. Thus, the offense should choose its goal line offense and the defense should respond with its goal line defense.

(c) Even if the defense chooses first, the offense will choose its goal line offense, so it does not help the defense to have the last move here.

31. **(a)**

Team at bat	Team in field	Team at bat	Outcome
	Pitcher X	Batter A	.320
		Batter B	.250
Batter A	Pitcher Y	Batter A	.240
		Batter B	.260
	Pitcher Z	Batter A	.280
		Batter B	.230
	Pitcher X		.250
Batter B	Pitcher Y		.260
	Pitcher Z		.230

(Start)

(b) The team at bat should leave Batter A up, and replace the batter with B only if the team in the field replaces the pitcher with Y, which it should do. If the team at bat brings in Batter B right away, the team in the field should switch to Pitcher Z.

33. **(a)** Because all three commissioners will vote their true preferences on the second vote, the possible outcomes of the second vote are residential over commercial 2 to 1, industrial over residential 2 to 1, and commercial over industrial 2 to 1. Herrera would vote against the Green bill on the first vote and for his own bill. The game tree follows.

Fong	Green	Fong	Outcome
	Yes on Green bill	Yes on Green bill	Commercial
Green bill first		No on Green bill	Industrial
	No on Green bill		Industrial
	Yes on Herrera bill		Commercial
Herrera bill first	No on Herrera bill	Yes on Herrera bill	Commercial
		No on Herrera bill	Residential

(Start)

(b) Backtracking through the tree, we see that Fong would vote yes on the Green bill and no on the Herrera bill. Thus, Green should vote yes on both bills. Whichever bill Fong selects to be first, the outcome will be commercial zoning.

35. We first eliminate all but the best of the Player's second moves, leaving the following partial game tree:

Player	Chance	Player	Chance	Outcome
	c	h		Player wins
a	d	k	s	Player wins
			t	Chance wins
	e	l	u	Chance wins
			v	Player wins
b	f	p		Player wins
	g	q	y	Player wins
			z	Chance wins

(Start)

By choosing a, the Player will win with probability $(1/3) + (1/3) \cdot (1/2) + (1/3) \cdot (1/2) = 2/3$. By choosing b, the Player will win with probability $(1/2) + (1/2) \cdot (1/2) = 3/4$. Therefore, the Player should choose b and then either p if Chance plays f and q if Chance plays g, with probability of winning 3/4.

37. (a)

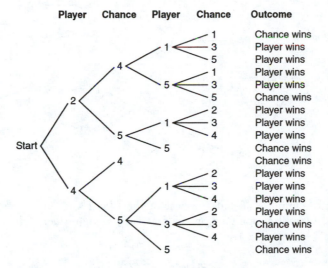

(b) We first eliminate all but the best of the Player's second moves (1 and 5 are equivalent after the Player moves to 2 and Chance to 4), leaving the following partial game tree:

By first moving to 2, the Player will win with probability $(1/2) \cdot (2 \cdot 3) + (1/2) = 5/6$. By first moving to 4, the Player will win with probability 1/2. Therefore, the Player should move first to 2 and then back to 1. The Player's probability of winning is 5/6.

39. **(a)**

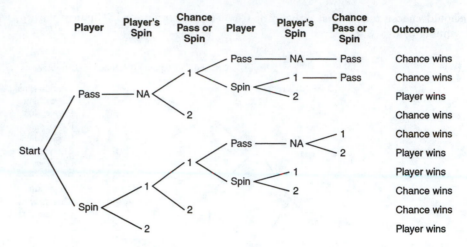

(b) We first eliminate all but the best of the Player's second turns (passing and spinning on the second turn after spinning on the first turn are equivalent), leaving the following partial game tree:

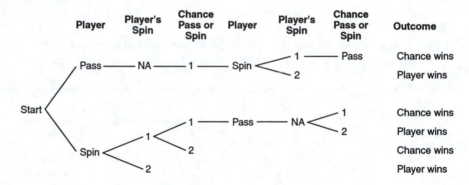

By passing on the first turn, the Player wins with probability 1/2. By spinning, the Player wins with probability $(1/2) \cdot (1/2) \cdot (1/2) + 1/2 = 5/8$. The Player should spin at the first opportunity and it doesn't matter on the second (if the game is still going).

(c) 5/8

SECTION 9.2

1. **(a)** (R2, C1)

(b) After the Column Player eliminates the dominated strategy C2, the Row Player will eliminate R1, leaving the reduced payoff matrix.

Column Player

		C1
Row Player	R2	(3, 4)

3. **(a)** (R1, C3) and (R3, C1)

(b) Strategy C2 is dominated by C3 (and C4). After its elimination, the Row Player should eliminate R2, leaving the reduced payoff matrix:

Column Player

	C1	C3	C4
R1	(6, 4)	(8, 6)	(7, 5)
R3	(8, 8)	(3, 3)	(5, 7)

Row Player

5. (a) none

   (b) Strategy R2 is dominated by R1. Its elimination yields the reduced payoff matrix.

Column Player

	C1	C2	C3
R1	(13, 0)	(5, 6)	(1, 8)
R3	(5, 12)	(3, 7)	(10, 4)

Row Player

7. (a) (R3, C2)

   (b) Strategies R2 and R4 are dominated by R1. Strategy C3 is dominated by C2. Following the elimination of all three strategies, the Column Player will eliminate C1 and C4. The Row Player will then eliminate R1, yielding the reduced payoff matrix.

Column Player

	C2
R3	(6, 6)

Row Player

9. (a)

Submissive Pig

	Push Lever	Do Not Push Lever
Push Lever	(7, 3)	(3, 8)
Do Not Push Lever	(12, –1)	(0, 0)

Dominant Pig

   (b) Each pig has a dominant strategy: the dominant pig should press the lever and the submissive pig should not.

11. (a)

Rock Hound #2

	6 Pieces	7 Pieces	8 Pieces	9 Pieces
6 Pieces	(150, 150)	(138, 161)	(126, 168)	(114, 171)
7 Pieces	(161, 138)	(147, 147)	(133, 152)	(119, 153)
8 Pieces	(168, 126)	(152, 133)	(136, 136)	(120, 135)
9 Pieces	(171, 114)	(153, 119)	(135, 120)	(117, 117)

Rock Hound #1

   (b) (8 pieces, 8 pieces)

   (c) The strategies of 6 and 7 pieces are dominated by 8 pieces for both rock hounds. After this reduction, both rock hounds will go on to eliminate 9 pieces, yielding the reduced payoff matrix.

Rock Hound #2

	8 Pieces
8 Pieces	(136, 136)

Rock Hound #1

   (d) Yes, both rock hounds have incentive to cheat because the agreed upon strategies of 6 pieces are dominated.

13. **(a)**

Rock Hound #2

	9 Pieces	10 Pieces	11 Pieces	12 Pieces
10 Pieces	(250, 225)	(230, 230)	(210, 231)	(190, 228)
11 Pieces	(253, 207)	(231, 210)	(209, 209)	(187, 204)
12 Pieces	(252, 189)	(228, 190)	(204, 187)	(180, 180)
13 Pieces	(247, 171)	(221, 170)	(195, 165)	(169, 156)

Rock Hound #1 labels the rows.

(b) (10 pieces, 11 pieces) and (11 pieces, 10 pieces)

(c) Strategies of 12 and 13 pieces for Rock Hound #1 are dominated by 11 pieces. The strategy of 12 pieces for Rock Hound #2 is dominated by both 10 pieces and 11 pieces. After this reduction, Rock Hound #2 will also eliminate 9 pieces, leaving the reduced payoff matrix.

Rock Hound #2

	10 Pieces	11 Pieces
10 Pieces	(230, 230)	(210, 231)
11 Pieces	(231, 210)	(209, 209)

Rock Hound #1 labels the rows.

(d) No. Both Rock Hounds have incentive to cheat if they could change from 10 to 11 pieces unilaterally. However, one or both would then be worse off.

15. **(a)**

Child

	Cooperative	Uncooperative
Mild Mannered	(4, 3)	(2, 4)
Harsh	(1, 1)	(3, 2)

Parent labels the rows.

(b) (harsh, uncooperative)

(c) For the child, being cooperative is dominated by being uncooperative. Thus, the parent will next eliminate the mild mannered approach.

Child

	Uncooperative
Harsh	(3, 2)

Parent labels the row.

17. **(a)**

Child

	Cooperative	Uncooperative
Mild Mannered	(4, 3)	(3, 4)
Harsh	(1, 1)	(2, 2)

Parent labels the rows.

(b) (mild mannered, uncooperative)

(c) For the parent, the harsh approach is dominated by the mild mannered approach. For the child, being cooperative is dominated by being uncooperative.

Child

	Uncooperative
Mild Mannered	(3, 4)

Parent labels the row.

19. (a)

	Seller	
Buyer	Do Not Negotiate	Lower Price If Asked
Buy	(3, 4)	(3, 4)
Negotiate and Go Elsewhere if Fail	(1, 1)	(4, 2)
Negotiate and Buy if Fail	(2, 3)	(4, 2)

(b) (buy, do not negotiate) and (negotiate and go elsewhere if fail, lower price if asked)

(c) For the Buyer, the negotiate and go elsewhere if fail strategy is dominated by the negotiate and buy if fail strategy. Once this strategy has been eliminated, the Seller will eliminate the strategy of lowering the price if asked. Finally, the Buyer will eliminate the negotiate and buy if fail strategy.

	Seller
Buyer	Do Not Negotiate
Buy	(3, 4)

21. (a)

	Seller	
Buyer	Do Not Negotiate	Lower Price If Asked
Buy	(2, 4)	(2, 4)
Negotiate and Go Elsewhere if Fail	(3, 1)	(4, 2)
Negotiate and Buy if Fail	(1, 3)	(4, 2)

(b) (negotiate and go elsewhere if fail, lower price if asked)

(c) The Buyer's strategy of negotiating and going elsewhere if it fails is dominant. After eliminating the Buyer's other strategies, the Seller will choose the strategy of lowering the price if asked.

	Seller
Buyer	Lower Price If Asked
Negotiate and Go Elsewhere If Fail	(4, 2)

23. The Buyer will play the game of Exercise 19 and buy without negotiation. The Seller would mistakenly be playing the game of Exercise 21, so would be willing to lower the price. The result will be that the Buyer will buy without negotiation.

25. (a) Write the bids as a pair in thousands of dollars, with the bid for the landscape first.

		Second Dealer					
		1 & 1	2 & 1	1 & 2	2 & 2	1 & 3	2 & 3
First Dealer	1 & 1	(2.5, 2.5)	(1.5, 2.5)	(1, 3)	(0, 3)	(1, 2)	(0, 2)
	2 & 1	(2.5, 1.5)	(2, 2)	(1, 2)	(0.5, 2.5)	(1, 1)	(0.5, 1.5)
	1 & 2	(3, 1)	(2, 1)	(2, 2)	(1, 2)	(1, 2)	(0, 2)
	2 & 2	(3, 0)	(2.5, 0.5)	(2, 1)	(1.5, 1.5)	(1, 1)	(0.5, 1.5)
	1 & 3	(2, 1)	(1, 1)	(2, 1)	(1, 1)	(1.5, 1.5)	(0.5, 1.5)

(b) (1 & 2, 1 & 2), (2 & 2, 2 & 2), (2 & 2, 2 & 3), (1 & 3, 1 & 3), (1 & 3, 2 & 3)

(c) The first dealer's strategies 1 &1, 2 & 1, and 1 & 2 are dominated by her strategy of 2 & 2. The second dealer's strategies 1 &1, 2 & 1, and 1 & 2 are dominated by her strategy of 2 & 2. Her strategy of 1 & 3 is dominated by her strategy of 2 & 3. After this partial reduction to a 2 × 2 game, the first dealer will eliminate the 1 & 3 strategy and the second dealer will eliminate the 2 & 2 strategy, and the game will be completely reduced.

	Second Dealer
	2 & 3
First Dealer 2 & 2	(0.5, 1.5)

(d) The first dealer should bid $2000 for each painting and the second $2000 for the landscape and $3000 for the portrait.

27. One example is

	Player B				
	C1	C2	C3	C4	C5
R1	(1,1)	(0,0)	(0,0)	(0,0)	(0,0)
R2	(0,0)	(1,1)	(0,0)	(0,0)	(0,0)
Player A R3	(0,0)	(0,0)	(1,1)	(0,0)	(0,0)
R4	(0,0)	(0,0)	(0,0)	(1,1)	(0,0)
R5	(0,0)	(0,0)	(0,0)	(0,0)	(1,1)

Another is

	Player B		
	C1	C2	C3
R1	(2,1)	(1,1)	(0,0)
Player A R2	(2,2)	(1,2)	(0,0)
R3	(0,0)	(0,0)	(1,1)

In both examples all nonzero payoffs occur at equilibrium points.

SECTION 9.3

1. (a) (R1, C2) and (R2, C2)

(b) Strategy C4 is dominated by C2. After its elimination, the Row Player will eliminate R1, yielding the reduced payoff matrix.

	Column Player		
	C1	C2	C3
R2	6	2	4
Row Player R3	−3	1	7
R4	7	0	−1

3. (a) (R2, C2)

(b) Strategy R2 is dominant and strategy C3 is dominated by C2. The final reduction is the elimination of C1.

	Column Player
	C2
Row Player R2	17

5. (a) none

(b) Strategy C2 is dominated by C1, and C3 is dominated by C4. Following this stage, R3 can be eliminated.

		Column Player	
		C1	C4
	R1	3.4	2.1
Row Player	R2	1.7	7.1
	R4	0	8.6

7. (a) none

(b) For the first player, betting with 2 only and not betting are both dominated by betting with an ace or 2. Similarly, for the second player, calling with a 2 only and folding with an ace or 2 are both dominated by calling with an ace or 2 and by calling with an ace only. No further reduction is possible.

		Second Player	
		Call with Ace or 2	Call with Ace Only
First Player	Bet with Ace or 2	0	$\frac{1}{7}$
	Bet with Ace Only	$\frac{2}{7}$	0

9. (a)

		Chaser		
		2	3	4
Runner	2	0	1	1
	4	1	2	0

(b) For the Chaser, moving to 3 is dominated by both of the other strategies.

		Chaser	
		2	4
Runner	2	0	1
	4	1	0

The Runner should choose between 2 and 4, as should the Chaser.

11. (a)

		Player B		
		C	D	H
	C	0	1	2
Player A	D	−1	0	1
	H	−2	−1	0

(b) (C, C)

(c) Strategy C is dominant for both players.

		Player B
		C
Player A	C	0

13. **(a)**

Player B

		C	D	H
	C	0	0	2
Player A	D	0	0	2
	H	−2	−2	0

(b) (C, C), (C, D), (D, C), (D, D)

(c) For both players, strategy H is dominated by their other two strategies.

Player B

		C	D
	C	0	0
Player A	D	0	0

15. **(a)**

Alex

		50	30	10
	60	0.2	0.2	0.2
Aaron	40	0	0.4	0.4
	20	0	−0.3	0.6

(b) For Alex, firing at 10 paces is dominated by both of his other strategies. Following its elimination, Aaron will not elect to fire at 20 paces. Next, Alex will eliminate firing at 30 paces. Finally, Aaron will eliminate firing at 40 paces.

Alex

		50
Aaron	60	0.2

(c) Aaron should shoot at 60 paces and Alex at 50.

(d) 0.6

17. If there are two equilibrium points in one column, the payoffs have to be the same or else the Row Player would prefer one to the other. Similarly, payoffs at equilibrium points in the same row must be the same or else the Column Player would prefer one to the other.

Now let $a_{i,j}$ denote the payoff for row i, column j and suppose row i, column j and row i', column j' are both equilibrium points with $i \neq i'$, $j \neq j'$. Because the Row Player does not want to switch strategy from row i, column j, we have $a_{i,j} \geq a_{i',j}$. Because the Column Player does not want to switch strategy from row i', column j', we have $a_{i',j} \geq a_{i',j'}$. Therefore, $a_{i,j} \geq a_{i',j'}$. However, the same reasoning implies $a_{i',j'} \geq a_{i,j}$, so that $a_{i,j} = a_{i',j'}$.

SECTION 9.4

1. **(a)** The expected payoffs for the Row Player's pure strategies are

$$R1: (0.4) \cdot 5 + (0.6) \cdot 3 = 3.8, \quad R2: (0.4) \cdot 3 + (0.6) \cdot 4 = 3.6.$$

The Row Player should choose R1.

(b) The expected payoffs for the Column Player's pure strategies are

$$C1: (0.7) \cdot 1 + (0.3) \cdot 4 = 1.9, \quad C2: (0.7) \cdot 2 + (0.3) \cdot 3 = 2.3.$$

The Column Player should choose C2.

(c) The expected payoff to the Row Player is $(0.7) \cdot (3.8) + (0.3) \cdot (3.6) = 3.74$. The expected payoff to the Column Player is $(0.4) \cdot (1.9) + (0.6) \cdot (2.3) = 2.14$.

3. (a) The expected payoffs for the Row Player's pure strategies are

$$R1: \tfrac{1}{3} \cdot 8 + \tfrac{2}{3} \cdot (-1) = 2, \quad R2: \tfrac{1}{3} \cdot 4 + \tfrac{2}{3} \cdot 5 = 4\tfrac{2}{3}.$$

The Row Player should choose R2.

(b) The expected payoffs for the Column Player's pure strategies are

$$C1: \tfrac{4}{5} \cdot 6 + \tfrac{1}{5} \cdot (-2) = 4\tfrac{2}{5}, \quad C2: \tfrac{4}{5} \cdot (-3) + \tfrac{1}{5} \cdot 7 = -1.$$

The Column Player should choose C1.

(c) The expected payoff to the Row Player is $\tfrac{4}{5} \cdot 2 + \tfrac{1}{5} \cdot \left(4\tfrac{2}{3}\right) = 2\tfrac{8}{15}$. The expected payoff to the Column Player is $\tfrac{1}{3} \cdot \left(4\tfrac{2}{5}\right) + \tfrac{2}{3} \cdot (-1) = \tfrac{4}{5}$.

5. (a) The expected payoffs for the Row Player's pure strategies are

$$R1: (0.4) \cdot 9 + (0.5) \cdot 5 + (0.1) \cdot 9 = 7, \quad R2: (0.4) \cdot (-2) + (0.5) \cdot 3 + (0.1) \cdot 7 = 1.4,$$
$$R3: (0.4) \cdot 1 + (0.5) \cdot 6 + (0.1) \cdot 5 = 3.9, \quad R4: (0.4) \cdot (-1) + (0.5) \cdot 5 + (0.1) \cdot 8 = 2.9.$$

The Row Player should choose R1.

(b) The expected payoffs for the Column Player's pure strategies are

$$C1: (0.1) \cdot 0 + (0.2) \cdot 8 + (0.4) \cdot (-4) + (0.3) \cdot 7 = 2.1,$$
$$C2: (0.1) \cdot 4 + (0.2) \cdot 5 + (0.4) \cdot 6 + (0.3) \cdot 8 = 6.2,$$
$$C3: (0.1) \cdot 1 + (0.2) \cdot 3 + (0.4) \cdot 5 + (0.3) \cdot 3 = 3.6.$$

The Column Player should choose C2.

(c) The expected payoff to the Row Player is $(0.1) \cdot 7 + (0.2) \cdot (1.4) + (0.4) \cdot (3.9) + (0.3) \cdot (2.9) = 3.41$. The expected payoff to the Column Player is $(0.4) \cdot (2.1) + (0.5) \cdot (6.2) + (0.1) \cdot (3.6) = 4.3$.

7. (a) The expected payoffs for the Row Player's pure strategies are

$$R1: \tfrac{2}{7} \cdot 2 + \tfrac{5}{7} \cdot 5 = 4\tfrac{1}{7}, \quad R2: \tfrac{2}{7} \cdot 7 + \tfrac{5}{7} \cdot 3 = 4\tfrac{1}{7}.$$

Because the two expected payoffs are the same, the Row Player may choose either strategy.

(b) The expected payoffs for the Column Player's pure strategies are

$$C1: \tfrac{7}{10} \cdot (-2) + \tfrac{3}{10} \cdot (-7) = -3\tfrac{1}{2}, \quad C2: \tfrac{7}{10} \cdot (-5) + \tfrac{3}{10} \cdot (-3) = -4\tfrac{2}{5}.$$

The Column Player should choose C1.

(c) The expected payoff to the Row Player is $4\tfrac{1}{7}$. The expected payoff to the Column Player is the opposite of this, or $-4\tfrac{1}{7}$.

9. (a) The expected payoffs for the Row Player's pure strategies are

$$R1: (0.2) \cdot 1 + (0.4) \cdot 3 + (0.1) \cdot 3 + (0.3) \cdot 1 = 2,$$
$$R2: (0.2) \cdot 2 + (0.4) \cdot (-1) + (0.1) \cdot 0 + (0.3) \cdot 2 = 0.6,$$
$$R3: (0.2) \cdot 3 + (0.4) \cdot 2 + (0.1) \cdot 3 + (0.3) \cdot (-1) = 1.4.$$

The Row Player should choose R1.

(b) The expected payoffs for the Column Player's pure strategies are

$$\text{C1:} \quad (0.2) \cdot (-1) + (0.3) \cdot (-2) + (0.5) \cdot (-3) = -2.3,$$
$$\text{C2:} \quad (0.2) \cdot (-3) + (0.3) \cdot 1 + (0.5) \cdot (-2) = -1.3,$$
$$\text{C3:} \quad (0.2) \cdot (-3) + (0.3) \cdot 0 + (0.5) \cdot (-3) = -2.1,$$
$$\text{C4:} \quad (0.2) \cdot (-1) + (0.3) \cdot (-2) + (0.5) \cdot 1 = -0.3.$$

The Column Player should choose C4.

(c) The expected payoff to the Row Player is $(0.2) \cdot 2 + (0.3) \cdot (0.6) + (0.5) \cdot (1.4) = 1.28$. The expected payoff to the Column Player is the opposite of this, or -1.28.

11. (a) $p = \dfrac{3 - 7}{2 - 5 - 7 + 3} = \dfrac{-4}{-7} = \dfrac{4}{7}, \quad q = \dfrac{3 - 5}{-7} = \dfrac{2}{7}$

The Row Player's optimal mixed strategy is to play R1 with probability 4/7 and R2 with probability 3/7. The Column Player's optimal mixed strategy is to play C1 with probability 2/7 and C2 with probability 5/7.

(b) The value of the game is $\dfrac{2 \cdot 3 - 5 \cdot 7}{-7} = 4\frac{1}{7}$.

13. (a) $p = \dfrac{(-3) - 7}{(-2) - 5 - 7 + (-3)} = \dfrac{-10}{-17} = \dfrac{10}{17}, \quad q = \dfrac{(-3) - 5}{-17} = \dfrac{8}{17}$

The Row Player's optimal mixed strategy is to play R1 with probability 10/17 and R2 with probability 7/17. The Column Player's optimal mixed strategy is to play C1 with probability 8/17 and C2 with probability 9/17.

(b) The value of the game is $\dfrac{(-2) \cdot (-3) - 5 \cdot 7}{-17} = 1\frac{12}{17}$.

15. (a) Strategy C2 is dominant, so should always be chosen by the Column Player. Thus, the Row Player should eliminate R1 and always play R2.

(b) The value of the game is the payoff at (R2, C2) or 0.4.

17. (a) Strategy R1 is dominated by R3. After the elimination of R1, the Column Player will eliminate C2, leaving the reduced payoff matrix.

Column Player

Row Player		C1	C3
	R2	1	-1
	R3	0	3

(b) $p = \dfrac{3 - 0}{1 - (-1) - 0 + 3} = \dfrac{3}{5}, \quad q = \dfrac{3 - (-1)}{5} = \dfrac{4}{5}$

The Row Player's optimal mixed strategy is to play R1 with probability 0, R2 with probability 3/5, and R3 with probability 2/5. The Column Player's optimal mixed strategy is to play C1 with probability 4/5, C2 with probability 0, and C3 with probability 1/5.

(c) The value of the game is $\dfrac{1 \cdot 3 - (-1) \cdot 0}{5} = \dfrac{3}{5}$.

19. (a) Strategy R2 is dominated by R4, C1 is dominated by C4, and C3 is dominated by C2. After the elimination of R4, C1, and C3, the Row Player will eliminate R3, leaving the reduced payoff matrix.

Column Player

Row Player		C2	C4
	R1	3	1
	R4	0	4

(b) $p = \dfrac{4-0}{3-1-0+4} = \dfrac{4}{6} = \dfrac{2}{3}, \quad q = \dfrac{4-1}{6} = \dfrac{1}{2}$

The Row Player's optimal mixed strategy is to play R1 with probability 2/3, R2 with probability 0, R3 with probability 0, and R4 with probability 1/3. The Column Player's optimal mixed strategy is to play C1 with probability 0, C2 with probability 1/2, C3 with probability 0, and C4 with probability 1/2.

(c) The value of the game is $\dfrac{3 \cdot 4 - 1 \cdot 0}{6} = 2.$

21. (a) The expected payoffs for the defendant's pure strategies are

$$\text{no lawyer: } (1/3) \cdot (-3000) + (2/3) \cdot (-3200) = -3133\tfrac{1}{3},$$

$$\text{hire a lawyer: } (1/3) \cdot (-800) + (2/3) \cdot (-3600) = -2666\tfrac{2}{3}.$$

The defendant should choose to hire a lawyer.

(b) The expected payoffs for the plaintiff's pure strategies are

$$\text{no lawyer: } (1/2) \cdot 3000 + (1/2) \cdot 0 = 1500,$$

$$\text{hire a lawyer: } (1/2) \cdot 2400 + (1/2) \cdot 2000 = 2200.$$

The plaintiff should choose to hire a lawyer.

(c) The expected payoff to the plaintiff is $(1/3) \cdot 1500 + (2/3) \cdot 2200 = \$1966\tfrac{2}{3}$. The expected payoff to the defendant is $(1/2) \cdot (-3133\tfrac{1}{3}) + (1/2) \cdot (-2666\tfrac{2}{3}) = -\$2900.$

23. (a) The expected payoffs for the first dealer's pure strategies are

$$\$100: \ (0.7) \cdot 150 + 0 \cdot 0 + (0.3) \cdot 0 = 105,$$

$$\$200: \ (0.7) \cdot 200 + 0 \cdot 100 + (0.3) \cdot 0 = 140,$$

$$\$300: \ (0.7) \cdot 100 + 0 \cdot 100 + (0.3) \cdot 50 = 85.$$

The first dealer should bid $200.

(b) The expected payoffs for the second dealer's pure strategies are

$$\$100: \ (0.1) \cdot 150 + (0.2) \cdot 0 + (0.7) \cdot 0 = 15,$$

$$\$200: \ (0.1) \cdot 200 + (0.2) \cdot 100 + (0.7) \cdot 0 = 40,$$

$$\$300: \ (0.1) \cdot 100 + (0.2) \cdot 100 + (0.7) \cdot 50 = 65.$$

The second dealer should bid $300.

(c) The expected payoff to the first dealer is $(0.1) \cdot 105 + (0.2) \cdot 140 + (0.7) \cdot 85 = \$98$. The expected payoff to the second dealer is $(0.7) \cdot 15 + 0 \cdot 40 + (0.3) \cdot 65 = \$30.$

25. (a) The expected payoffs for the Row Player's pure strategies are

$$\text{rock: } (0.7) \cdot 0 + (0.2) \cdot (-1) + (0.1) \cdot 1 = -0.1,$$

$$\text{paper: } (0.7) \cdot 1 + (0.2) \cdot 0 + (0.1) \cdot (-1) = 0.6,$$

$$\text{scissors: } (0.7) \cdot (-1) + (0.2) \cdot 1 + (0.1) \cdot 0 = -0.5.$$

The Row Player should choose paper.

(b) The expected payoffs for the Column Player's pure strategies are

$$\text{rock: } (0.2) \cdot 0 + (0.4) \cdot (-1) + (0.4) \cdot 1 = 0,$$
$$\text{paper: } (0.2) \cdot 1 + (0.4) \cdot 0 + (0.4) \cdot (-1) = -0.2,$$
$$\text{scissors: } (0.2) \cdot (-1) + (0.4) \cdot 1 + (0.4) \cdot 0 = 0.2.$$

The Column Player should choose scissors.

(c) The expected payoff to the Row Player is $(0.2) \cdot (-0.1) + (0.4) \cdot (0.6) + (0.4) \cdot (-0.5) = 0.02$. The expected payoff to the Column Player is the opposite of this, or -0.02.

27. Your expected payoffs are $(0.6) \cdot 1700 + (0.3) \cdot 1200 + (0.1) \cdot 900 = \$1470$ for stocks and $\$1200$ for a CD. You should invest in stocks.

29. (a)

		Second Player	
		One Finger	Two Fingers
First Player	One Finger	1	−1
	Two Fingers	−1	1

(b) The first player's expected values are

$$\text{one finger: } (3/5) \cdot 1 + (2/5) \cdot (-1) = 1/5,$$
$$\text{two fingers: } (3/5) \cdot (-1) + (2/5) \cdot 1 = -1/5.$$

The first player should choose one finger.

(c) The expected payoff to the first player is $(2/3) \cdot (1/5) + (1/3) \cdot (-1/5) = 1/15$. The expected payoff to the second player is the opposite of this, or $-1/15$.

(d) $p = \dfrac{1 - (-1)}{1 - (-1) - (-1) + 1} = \dfrac{2}{4} = \dfrac{1}{2}, \quad q = \dfrac{1 - (-1)}{1 - (-1) - (-1) + 1} = \dfrac{1}{2}$

Both players should choose one finger with probability 1/2 and two fingers with probability 1/2.

(e) The value of the game is $v = \dfrac{1 \cdot 1 - (-1) \cdot (-1)}{4} = 0.$

31. (a)

		Receiver	
		Guess Forehand	Guess Backhand
Server	Serve to Forehand	40	70
	Serve to Backhand	72	54

(b) $p = \dfrac{54 - 72}{40 - 70 - 72 + 54} = \dfrac{-18}{-48} = \dfrac{3}{8}, \quad q = \dfrac{54 - 70}{-48} = \dfrac{1}{3}$

The server should serve to the forehand with probability 3/8 and to backhand with probability 5/8. The receiver should guess forehand with probability 1/3 and guess backhand with probability 2/3.

(c) The value of the game is $v = \dfrac{40 \cdot 54 - 70 \cdot 72}{-48} = 60$, i.e. the server should win 60% of the points.

33. (a)

		Receiver	
		Guess Forehand	Guess Backhand
Server	Serve to Forehand	40	70
	Serve to Backhand	90	75

(b) Serving to the backhand is a dominant strategy. The receiver should then never guess forehand. Therefore, the server should serve to the forehand with probability 0 and to backhand with probability 1. The receiver should guess forehand with probability 0 and guess backhand with probability 1.

(c) The value of the game is 75, i.e. the server should win 75% of the points.

35. (a)

		von Kluge	
		Attack Gap	Withdraw
	Reinforce Gap	2	3
Bradley	Move Eastward	1	5
	Hold in Reserve	6	4

(b) Bradley's strategy of reinforcing the gap is dominated by holding the army in reserve. No further reduction is possible.

		von Kluge	
		Attack Gap	Withdraw
	Move Eastward	1	5
Bradley	Hold in Reserve	6	4

(c) $p = \dfrac{4-6}{1-5-6+4} = \dfrac{-2}{-6} = \dfrac{1}{3}, \quad q = \dfrac{4-5}{-6} = \dfrac{1}{6}$

Bradley's optimal mixed strategy is to reinforce the gap with probability 0, to move eastward with probability 1/3, and to hold with probability 2/3. Von Kluge's optimal mixed strategy is to attack with probability 1/6 and to withdraw with probability 5/6.

(d) The value of the game is $\dfrac{1 \cdot 4 - 5 \cdot 6}{-6} = 4\frac{1}{3}$, slightly better for the Allies than the commanders' initial decisions.

37. (a)

		von Kluge	
		Attack Gap	Withdraw
	Reinforce Gap	0	1
Bradley	Move Eastward	−5	7
	Hold in Reserve	10	3

(b) Bradley's strategy of reinforcing the gap is dominated by holding the army in reserve. No further reduction is possible.

		von Kluge	
		Attack Gap	Withdraw
	Move Eastward	−5	7
Bradley	Hold in Reserve	10	3

(c) $p = \dfrac{3-10}{(-5)-7-10+3} = \dfrac{-7}{-19} = \dfrac{7}{19}, \quad q = \dfrac{3-7}{-19} = \dfrac{4}{19}$

Bradley's optimal mixed strategy is to reinforce the gap with probability 0, to move eastward with probability 7/19, and to hold with probability 12/19. Von Kluge's optimal mixed strategy is to attack with probability 4/19 and to withdraw with probability 15/19.

(d) The value of the game is $\dfrac{(-5) \cdot 3 - 7 \cdot 10}{-19} = 4\frac{9}{19}$, somewhat better for the Allies than the commanders' initial decisions.

39. (a)

		IRS	
		Check First Line	Check Second Line
Cheat	Cheat on First Line	−2000	1000
	Cheat on Second Line	4000	−2000

(b) $p = \dfrac{(-2000) - 4000}{(-2000) - 1000 - 4000 + (-2000)} = \dfrac{-6000}{-9000} = \dfrac{2}{3}, \quad q = \dfrac{(-2000) - 1000}{-9000} = \dfrac{1}{3}$

The cheat should cheat on the first line with probability 2/3 and cheat on the second line with probability 1/3. The IRS should check the first line with probability 1/3 and check the second line with probability 2/3.

(c) The value of the game is $\dfrac{(-2000) \cdot (-2000) - 1000 \cdot 4000}{-9000} = 0$, so the cheat's expected value is $0.

41. (a)

		IRS	
		Check First Line	Check Second Line
Cheat	Cheat on First Line	−2000	1000
	Cheat on Second Line	4000	−8000

(b) $p = \dfrac{(-8000) - 4000}{(-2000) - 1000 - 4000 + (-8000)} = \dfrac{-12{,}000}{-15{,}000} = \dfrac{4}{5}, \quad q = \dfrac{(-8000) - 1000}{-15{,}000} = \dfrac{3}{5}$

The cheat should cheat on the first line with probability 4/5 and cheat on the second line with probability 1/5. The IRS should check the first line with probability 3/5 and check the second line with probability 2/5.

(c) The value of the game is $\dfrac{(-2000) \cdot (-8000) - 1000 \cdot 4000}{-15{,}000} = -800$, so the cheat's expected value is an $800 loss.

43. (a)

		IRS	
		Check First Line	Check Second Line
Cheat	Cheat on First Line	−5000	1000
	Cheat on Second Line	4000	−20,000

(b) $p = \dfrac{(-20{,}000) - 4000}{(-5000) - 1000 - 4000 + (-20{,}000)} = \dfrac{-24{,}000}{-30{,}000} = \dfrac{4}{5}, \quad q = \dfrac{(-20{,}000) - 1000}{-30{,}000} = \dfrac{7}{10}$

The cheat should cheat on the first line with probability 4/5 and cheat on the second line with probability 1/5. The IRS should check the first line with probability 7/10 and check the second line with probability 3/10.

(c) The value of the game is $\dfrac{(-5000) \cdot (-20{,}000) - 1000 \cdot 4000}{-30{,}000} = -3200$, so the cheat's expected value is a $3200 loss.

45.

		Sheriff	
		North	South
Robber	North	2/3	1
	South	1	3/4

$$p = \dfrac{\frac{3}{4} - 1}{\frac{2}{3} - 1 - 1 + \frac{3}{4}} = \dfrac{-\frac{1}{4}}{-\frac{7}{12}} = \dfrac{3}{7}, \quad q = \dfrac{\frac{3}{4} - 1}{-\frac{7}{12}} = \dfrac{3}{7}$$

The robber should go north with probability 3/7 and south with probability 4/7. The sheriff should go north with probability 3/7 and south with probability 4/7. The probability of escape is $v =$
$$\frac{\frac{2}{3} \cdot \frac{3}{4} - 1 \cdot 1}{-\frac{7}{12}} = \frac{6}{7}.$$

47.

		Goalie	
		Guess Left	Guess Right
Kicker	Kick Left	50	90
	Kick Right	90	50

$$p = \frac{50 - 90}{50 - 90 - 90 + 50} = \frac{-40}{-80} = \frac{1}{2}, \quad q = \frac{50 - 90}{-80} = \frac{1}{2}$$

The kicker should kick left with probability 1/2 and kick right with probability 1/2. The goalie should guess left with probability 1/2 and guess right with probability 1/2. The probability of a goal is
$$v = \frac{50 \cdot 50 - 90 \cdot 90}{-80} = 70, \text{ or } 70\%.$$

49.

		Goalie	
		Guess Left	Guess Right
Kicker	Kick Left	50	95
	Kick Right	85	50

$$p = \frac{50 - 85}{50 - 95 - 85 + 50} = \frac{-35}{-80} = \frac{7}{16}, \quad q = \frac{50 - 95}{-80} = \frac{9}{16}$$

The kicker should kick left with probability 7/16 and kick right with probability 9/16. The goalie should guess left with probability 9/16 and guess right with probability 7/16. The probability of a goal is $v = \dfrac{50 \cdot 50 - 95 \cdot 85}{-80} = 69.6875$, or 69.6875%.

SECTION 9.5

1.

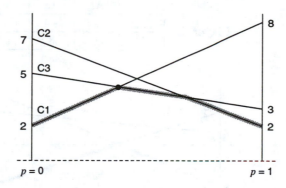

The Column Player will play a mixture of C1 and C3.

$$p = \frac{5 - 2}{8 - 3 - 2 + 5} = \frac{3}{8}, \quad q = \frac{5 - 3}{8} = \frac{1}{4}$$

The Row Player should play R1 with probability 3/8 and R2 with probability 5/8. The Column Player should play C1 with probability 1/4, C2 with probability 0, and C3 with probability 3/4.

3.

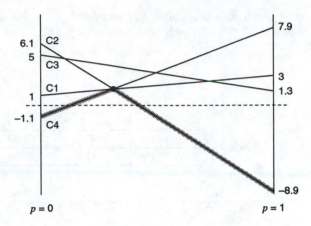

Three lines intersect at a single point. The Column Player will play a mixture of the two outside lines — C2 and C4.

$$p = \frac{-1.1 - 6.1}{(-8.9) - 7.9 - 6.1 + (-1.1)} = \frac{-7.2}{-24} = \frac{3}{10}, \quad q = \frac{-1.1 - 7.9}{-24} = \frac{3}{8}$$

The Row Player should play R1 with probability 3/10 and R2 with probability 7/10. The Column Player should play C1 with probability 0, C2 with probability 3/8, C3 with probability 0, and C4 with probability 5/8.

5. Strategies A2, B2, and B3 are all dominated and no further reduction is possible.

		Player B	
		B1	B4
	A1	−3	−1
Player A	A3	4	−3
	A4	−5	4

We rewrite the game with Player B as the row player so that the row player has only two strategies.

		Player A		
		A1	A3	A4
Player B	B1	3	−4	5
	B4	1	3	−4

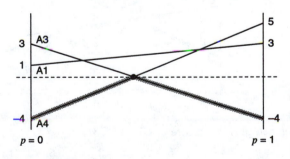

Player A will play a mixture of A3 and A4.

$$p = \frac{(-4) - 3}{(-4) - 5 - 3 + (-4)} = \frac{-7}{-16} = \frac{7}{16}, \quad q = \frac{(-4) - 5}{-16} = \frac{9}{16}$$

Player A will play A1 with probability 0, A2 with probability 0, A3 with probability 9/16, and A4 with probability 7/16. Player B will play B1 with probability 7/16, B2 with probability 0, B3 with probability 0, and B4 with probability 9/16.

7. Strategies A2 and A4 are dominated. Following their elimination, B3 may be eliminated.

Player B

		B1	B2	B4	B5
Player A	A1	7	−19	−7	48
	A3	−36	44	−21	−47

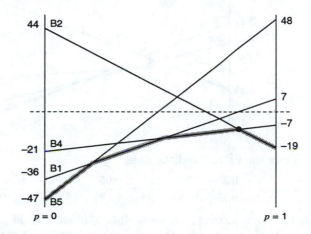

Player B will play a mixture of B2 and B4.

$$p = \frac{(-21) - 44}{(-19) - (-7) - 44 + (-21)} = \frac{-65}{-77} = \frac{65}{77}, \quad q = \frac{(-21) - (-7)}{-77} = \frac{2}{11}$$

Player A will play A1 with probability 65/77, A2 with probability 0, A3 with probability 12/77, and A4 with probability 0. Player B will play B1 with probability 0, B2 with probability 2/11, B3 with probability 0, B4 with probability 9/11, and B5 with probability 0.

9. (a)

Second Player

		One Finger	Two Fingers	Three Fingers
First Player	One Finger	2	−3	4
	Two Fingers	−3	4	−5

(b)   The matrix does not reduce, so we draw the strategy lines.

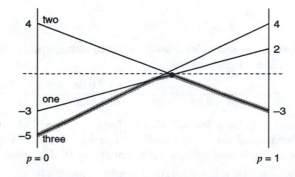

From the graph, we see that the second player will play a mixture of one and two fingers.

$$p = \frac{4-(-3)}{2-(-3)-(-3)+4} = \frac{7}{12}, \quad q = \frac{4-(-3)}{12} = \frac{7}{12}$$

The First Player will choose one finger with probability 7/12 and two fingers with probability 5/12. The Second Player will choose one finger with probability 7/12, two fingers with probability 5/12, and three fingers with probability 0.

(c) The value of the game is $\dfrac{2 \cdot 4 - (-3) \cdot (-3)}{12} = -\dfrac{1}{12}$.

11.

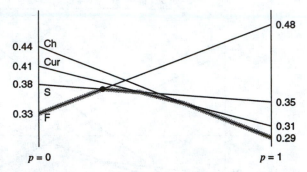

The pitcher should mix the fastball and the slider.

$$p = \frac{0.38 - 0.33}{0.48 - 0.35 - 0.33 + 0.38} = \frac{0.05}{0.18} = \frac{5}{18}, \quad q = \frac{0.38 - 0.35}{0.18} = \frac{1}{6}$$

The batter should look for a fastball with probability 5/18 and should not anticipate with probability 13/18. The pitcher should throw a fastball with probability 1/6, a change-up with probability 0, a curve with probability 0, and a slider with probability 5/6. The value of the game is

$$\frac{0.48 \cdot 0.38 - 0.35 \cdot 0.33}{0.18} = \frac{0.0669}{0.18} = \frac{223}{600} \approx 0.372.$$

13.

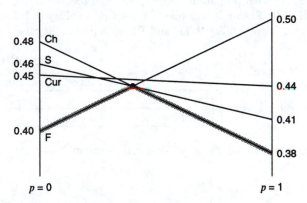

The fastball, change-up, and slider lines intersect in a single point. The pitcher should mix the outside two — the fastball and the change-up.

$$p = \frac{0.48 - 0.40}{0.50 - 0.38 - 0.40 + 0.48} = \frac{0.08}{0.20} = \frac{2}{5}, \quad q = \frac{0.48 - 0.38}{0.20} = \frac{1}{2}$$

The batter should look for a fastball with probability 2/5 and should not anticipate with probability 3/5. The pitcher should throw a fastball with probability 1/2, a change-up with probability 1/2, a curve with probability 0, and a slider with probability 0. The value of the game is

$$\frac{0.50 \cdot 0.48 - 0.38 \cdot 0.40}{0.20} = \frac{11}{25} = 0.44.$$

15. **(a)**

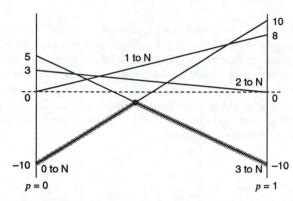

		Defender			
		0 to N	1 to N	2 to N	3 to N
Attacker	Attack N	10	8	0	−10
	Attack S	−10	0	3	5

(b)

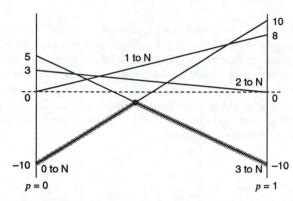

The Defender should mix 0 to N and 3 to N.

$$p = \frac{5 - (-10)}{10 - (-10) - (-10) + 5} = \frac{15}{35} = \frac{3}{7}, \quad q = \frac{5 - (-10)}{35} = \frac{3}{7}$$

The Attacker should attack Northridge with probability 3/7 and attack Southport with probability 4/7. The Defender should allocate 0 to Northridge with probability 3/7, 1 to Northridge with probability 0, 2 to Northridge with probability 0, and 3 to Northridge with probability 4/7.

(c) The value of the game is $\frac{10 \cdot 5 - (-10) \cdot (-10)}{35} = \frac{-50}{35} = -1\frac{3}{7}$. The attacker should not invade the defending country.

17. **(a)**

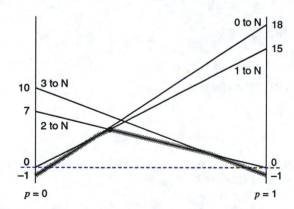

		Defender			
		0 to N	1 to N	2 to N	3 to N
Attacker	Attack N	18	15	0	−1
	Attack S	−1	0	7	10

(b)

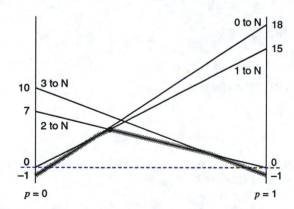

The Defender should mix 1 to N and 2 to N.

$$p = \frac{7 - 0}{15 - 0 - 0 + 7} = \frac{7}{22}, \quad q = \frac{7 - 0}{22} = \frac{7}{22}$$

The Attacker should attack Northridge with probability 7/22 and attack Southport with probability 15/22. The Defender should allocate 0 to Northridge with probability 0, 1 to Northridge with probability 7/22, 2 to Northridge with probability 15/22, and 3 to Northridge with probability 0.

(c) The value of the game is $\dfrac{15 \cdot 7 - 0 \cdot 0}{22} = \dfrac{105}{22} = 4\frac{17}{22}$. The attacker should invade the defending country.

19. The strategy line for the dominated strategy should be completely above the strategy line that dominates it.

21. The expected payoff to the Row Player against any of the Column Player's strategies has the form $pa + (1-p)b$. Any mixture of two such lines will have the same form. Any mixture of the two outside strategies must pass through the common point of intersection with the middle strategy. If we choose a mixture whose expected payoff agrees with that of the middle strategy for $p = 0$, then the line determined by this mixture passes through two points of the line for the middle strategy, hence must coincide with it. The Column Player can replace the middle strategy with a mixture of the outside strategies in a way that will obtain exactly the same expected payoff for any mixture of the two strategies the Row Player can choose.

SECTION 9.6

1. Some possibilities are January 1, July 4, October 31, and December 25.

3. George Washington and Abraham Lincoln

5. Two possibilities are pepperoni and sausage.

7. (a) R1: C3, R2: C2, R3: C2. Strategies R1 and R3 are credible.

   (b) C1: R3, C2: R3, C3: R1. Strategies C2 and C3 are credible.

9. (a) R1: C1, R2: C4, R3: C1. Strategy R3 is credible.

   (b) C1: R3, C2: R2, C3: R1, C4: R1. Strategy C1 is credible.

11. (a) Row Player: 4, Column Player: 8

    (b) (R1, C2), (R1, C3), (R2, C2), (R3, C2)

    (c) (R1, C3)

    (d) (R1, C3), (R3, C2)

13. (a) Row Player: 2, Column Player: 6

    (b) (R1, C2), (R1, C3), (R2, C2)

    (c) (R1, C3), (R2, C2)

    (d) (R2, C2)

15. We will let R, MV, and S stand for Riverside, Mountain View, and Sunnyvale, respectively. We express our payoffs in terms of thousands of people.

    (a)

		Post		
		R	MV	S
	R	(30, 10)	(40, 30)	(40, 20)
Times	MV	(30, 40)	(22.5, 7.5)	(30, 20)
	S	(20, 40)	(20, 30)	(15, 5)

(b)

	Post
	MV
Times R	(40, 30)

(c) Times: R, MV; Post: R, MV

(d) Times: 30, Post: 10

(e) (R, R), (R, MV), (R, S), (MV, R), (MV, S)

(f) (R, MV), (MV, R)

(g) (R, MV), (MV, R)

17. We will let R, MV, and S stand for Riverside, Mountain View, and Sunnyvale, respectively. We express our payoffs in terms of thousands of people.

(a)

	Post		
	R	MV	S
R & MV	(50, 20)	(55, 15)	(70, 20)
Times R & S	(40, 20)	(60, 30)	(50, 10)
MV & S	(50, 40)	(35, 15)	(40, 10)

(b)

	Post	
	R	MV
R & MV	(50, 20)	(55, 15)
Times R & S	(40, 20)	(60, 30)

(c) Times: all, Post: all

(d) Times: 50, Post: 20

(e) (R & MV, R), (R & MV, S), (R & S, MV), (MV & S, R)

(f) (R & MV, S), (R & S, MV), (MV & S, R)

(g) (R & MV, R), (R & MV, S), (R & S, MV), (MV & S, R)

19. **(a)** We write payoffs in units of $1000.

		Defendant		
		No Lawyer	Mediocre Lawyer	Good Lawyer
	No Lawyer	(10, −10)	(0, −2)	(0, −4)
Plaintiff	Mediocre Lawyer	(8, −10)	(8, −12)	(−2, −4)
	Good Lawyer	(6, −10)	(6, −12)	(6, −14)

(b) We write payoffs in units of $1000.

		Defendant		
		No Lawyer	Mediocre Lawyer	Good Lawyer
	No Lawyer	(10, −10)	(0, −2)	(0, −4)
Plaintiff	Mediocre Lawyer	(8, −10)	(8, −12)	(−2, −4)
	Good Lawyer	(6, −10)	(6, −12)	(6, −14)

(c) Plaintiff: 6, Defendant: −10

(d) (no lawyer, no lawyer), (mediocre lawyer, no lawyer), (good lawyer, no lawyer)

(e) (no lawyer, no lawyer)

(f) none

21. (a) We write payoffs in units of $1000.

		Defendant		
		No Lawyer	Mediocre Lawyer	Good Lawyer
Plaintiff	No Lawyer	(2, −2)	(0, −2)	(0, −4)
	Mediocre Lawyer	(2, −4)	(0, −4)	(−2, −4)
	Good Lawyer	(0, −4)	(0, −6)	(−2, −6)

(b)

		Defendant
		No Lawyer
Plaintiff	No Lawyer	(2, −2)

(c) Plaintiff: 0, Defendant: −4

(d) (no lawyer, no lawyer), (no lawyer, mediocre lawyer), (no lawyer, good lawyer), (mediocre lawyer, no lawyer), (mediocre lawyer, mediocre lawyer), (good lawyer, no lawyer)

(e) (no lawyer, no lawyer)

(f) (no lawyer, no lawyer), (no lawyer, mediocre lawyer), (mediocre lawyer, no lawyer), (mediocre lawyer, mediocre lawyer)

23. (a)

		Country 2		
		No Buildup	Moderate Buildup	Large Buildup
Country 1	No Buildup	(5, 5)	(2, 7)	(2, 6)
	Moderate Buildup	(7, 2)	(4, 4)	(1, 6)
	Large Buildup	(6, 2)	(6, 1)	(3, 3)

(b)

		Country 2
		Large Buildup
Country 1	Large Buildup	(3, 3)

(c) Country 1: 3, Country 2: 3

(d) (no buildup, no buildup), (moderate buildup, moderate buildup), (large buildup, large buildup)

(e) (no buildup, no buildup)

(f) (large buildup, large buildup)

25. (a)

		Country 2		
		No Buildup	Moderate Buildup	Large Buildup
Country 1	No Buildup	(3, 5)	(3, 4)	(3, 3)
	Moderate Buildup	(5, 2)	(2, 4)	(2, 3)
	Large Buildup	(4, 2)	(4, 1)	(1, 3)

(b)

	Country 2		
	No Buildup	Moderate Buildup	Large Buildup
Country 1 No Buildup	(3, 5)	(3, 4)	(3, 3)
Moderate Buildup	(5, 2)	(2, 4)	(2, 3)
Large Buildup	(4, 2)	(4, 1)	(1, 3)

(c) Country 1: 3, Country 2: 3

(d) (no buildup, no buildup), (no buildup, moderate buildup), (no buildup, large buildup)

(e) (no buildup, no buildup)

(f) none

27. (a)

	Politician 2			
	For A & B	For A	For B	For Neither
Politician 1 For A & B	(3, 3)	(4, 1)	(1, 4)	(2, 2)
For A	(4, 1)	(4, 1)	(2, 2)	(2, 2)
For B	(1, 4)	(2, 2)	(1, 4)	(2, 2)
For Neither	(2, 2)	(2, 2)	(2, 2)	(2, 2)

(b) AB

29. (a)

	Politician 2			
	For A & B	For A	For B	For None
Politician 1 For C	(7, 5)	(4, 7)	(3, 6)	(1, 8)
For None	(8, 1)	(6, 3)	(5, 2)	(2, 4)

(b) ABC, AC, BC

31. (a)

	Politician 2			
	For A & B	For A	For B	For None
Politician 1 For C & D	(11, 8)	(8, 10)	(3, 15)	(1, 16)
For C	(15, 3)	(13, 5)	(7, 11)	(5, 13)
For D	(12, 4)	(10, 6)	(4, 12)	(2, 14)
For None	(16, 1)	(14, 2)	(9, 7)	(6, 9)

(b) ACD, BC

33. Whenever a player has payoff less than his or her security level, the player would want to unilaterally switch strategies to one that guarantees the security level. Thus, if a point has payoff less than the player's security level, the point cannot be an equilibrium point.

Chapter 9 Review Exercises

1.

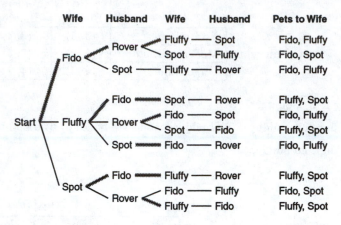

Highlighting the best choices at each juncture, choosing arbitrarily when results are the same, we see that the wife should choose Fido, the husband should then choose Rover, the wife should then choose Fluffy, leaving Spot for the husband. (We have omitted highlights for the husband's last pick, since he has no options.)

3. In constructing the game tree, note that B's responses of 1, 2 and 5, 6 to 3, 4 are strategically the same, so have been compressed into one branch.

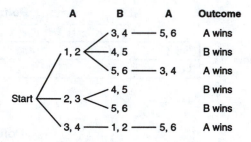

By choosing squares 3 and 4, Player A will win.

5. By covering squares 1 and 2, the first player becomes the second to move on a board with $n - 2$ squares. By covering squares 2 and 3, neither player can cover square 1 and the game becomes one on $n - 3$ squares with the original first player moving second. Thus, if the second player has a winning strategy in one of the two cases, the first player will win the game with n squares under best play.

7. (a)

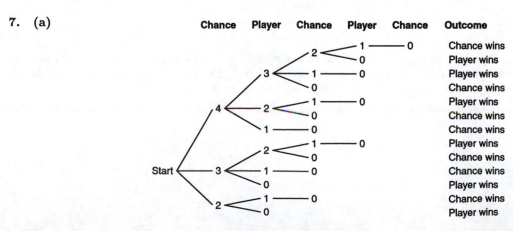

(b) From 2 or 3 left after Chance's first move, the Player should win. From 4 left, the Player should leave 3, and then win if possible. The Player's probability of winning is $(1/3) \cdot (2/3) + (1/3) + (1/3) = 8/9$.

9. (a)

		Runner		
		2, 1	3, 1	3, 2
	2, 3	2	−2	1
Chaser	2, 4	2	−2	−2
	3, 4	−2	2	2

(b) For the Chaser, strategy 2, 4 is dominated. For the Runner, strategy 3, 2 is dominated. The reduced matrix is

		Runner	
		2, 1	3, 1
	2, 3	2	−2
Chaser	3, 4	−2	2

$$p = \frac{2 - (-2)}{2 - (-2) - (-2) + 2} = \frac{4}{8} = \frac{1}{2}, \quad q = \frac{2 - (-2)}{8} = \frac{1}{2}$$

The Chaser should elect 2, 3 and 3, 4 each with probability 1/2, and 2,4 with probability 0. The Runner should elect 2, 1 and 3, 1 each with probability 1/2, and 3, 2 with probability 0.

(c) The value of the game is $\dfrac{2 \cdot 2 - (-2) \cdot (-2)}{8} = 0$.

11. The dominated strategies are 4, 5, and 6 for the Column Player. Following their elimination, the Row Player will eliminate 2 and 3. Next the Column Player will eliminate 1, and then the Row Player will eliminate 1, leaving the reduced matrix.

		Column Player		
		0	2	3
Row Player	0	0	225	450
	4	500	325	150

We plot the three strategy lines.

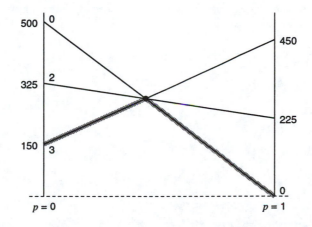

The three lines intersect in a single point. The Column Player's optimal strategy will mix the outside two — 0 and 3.

$$p = \frac{150 - 500}{0 - 450 - 500 + 150} = \frac{-350}{-800} = \frac{7}{16}, \quad q = \frac{150 - 450}{-800} = \frac{3}{8}$$

The Row Player should choose 0 with probability 7/16, 1 with probability 0, 2 with probability 0, 3 with probability 0, and 4 with probability 9/16. The Column Player should choose 0 with probability 3/8, 1 with probability 0, 2 with probability 0, 3 with probability 5/8, 4 with probability 0, 5 with probability 0, and 6 with probability 0.

13. (a)

Column Fool

		Swerve	Do Not Swerve
Row Fool	Swerve	(3, 3)	(2, 4)
	Do Not Swerve	(4, 2)	(1, 1)
	Swerve Late	(4, 2)	(2, 4)

(b) Swerving late is the dominant strategy for the Row Fool. Therefore the Row Fool should swerve late; the Column Fool will then choose not to swerve.

15. (a) Let the strategy be the number grabbed.

Player B

		0	1	2	3	4
	0	(8, 8)	(6, 7)	(4, 6)	(2, 5)	(0, 4)
	1	(7, 6)	(5, 5)	(3, 4)	(1, 3)	(1, 3)
Player A	2	(6, 4)	(4, 3)	(2, 2)	(2, 2)	(2, 2)
	3	(5, 2)	(3, 1)	(2, 2)	(2, 2)	(2, 2)
	4	(4, 0)	(3, 1)	(2, 2)	(2, 2)	(2, 2)

(b) For each player, grabbing 3 or 4 nuts are dominated strategies. Following their elimination, both players should eliminate grabbing 1 or 2 nuts. Therefore, neither player should grab any nuts.

(c) The security levels are 2 for each, with everything in the negotiation set except the strategy pairs (0, 4), (1, 3), (1, 4), (3, 1), (4, 0), and (4, 1). The optimal negotiation set is for each to grab 0. The stable points are when both try to grab 3 or 4 and when both grab 0.

17. the left and middle squares

Notes

Notes

Notes

Notes

Notes

Notes

Notes

Notes

Notes

Notes

Notes